T0293663

Radiation Biology for Medical Physicists

Radiation Biology for Medical Physicists

C.S. Sureka and C. Armpilia

CRC Press is an imprint of the
Taylor & Francis Group, an **informa** business

CRC Press
Taylor & Francis Group
6000 Broken Sound Parkway NW, Suite 300
Boca Raton, FL 33487-2742

Printed on acid-free paper

International Standard Book Number-13: 978-1-4987-6589-3 (Hardback)

Visit the Taylor & Francis Web site at
http://www.taylorandfrancis.com

and the CRC Press Web site at
http://www.crcpress.com

Contents

Preface

Radiation Biology for Medical Physicists is written based on our experience in teaching a radiobiology paper to postgraduate medical physics students. This book has 10 chapters. The first nine are structured for a 16-week, one-semester paper after completing the "Fundamental Radiation Physics" and "Anatomy and Physiology as Applied to Oncology and Imaging" papers. These nine chapters target clinical medical physicists and medical physics students as they discuss more about the importance of radiobiological models in radiation therapy. The last chapter is on "biological dosimetry" to encourage medical physics students to choose radiobiological research as their career and clinical medical physicists to perform radiobiological research as part of their job. Many self-drawing figures and figures taken from various sources have been added to pictorially explain the concepts. Multiple-choice questions are given at the end of each chapter to test the knowledge. All this was possible because of the support of the International Centre for Theoretical Physics (ICTP), Italy, through the ICTP Junior Associateship scheme. Last but not the least, We express our heartfelt thanks to all of our students, seniors, and contributors for their suggestions.

Acknowledgments

We are very thankful to many colleagues, professionals, and students for their suggestions and, sometimes, extra duties. The manuscript took around three years to mature and be written down. During this period, we interacted with so many people who understood the importance of this work and firmly supported us. Special thanks go to Colin G. Orton, K. Moorthy, R. Dale, R. K. Jeevan Ram, G. Ramanathan, S. Ganesan, P. Aruna, Arun Chougule, P. Kolandaivel, Luciano Bertocchi, Cheng B Saw, A. Gopiraj, N. Amuthavalli, T. Sundaram, P. Venkatraman, S. Sangeetha, S. K. Grace Mercy, C. Senthamilselvan, K. Mayakannan, K. Indhumathi, A. Jagadeeswari, K. Muthumari, S. Prakash, and P. Premkumar, to whom we offer our utmost recognition and respect. We also thank the staff at CRC Press for their full dedication to this project; they encouraged us and filled us with enthusiasm. We particularly thank Aastha Sharma, Shikha Garg, Edwards Alex, Anto Aroshini Raphael, and Rachael Panthier for their special support in moments we really needed help. Thanks to our families and friends, who were deprived of our company and support for many months. Thanks to the Almighty for making this proposal a successful one. Finally, we express our gratitude to our institutes—Bharathiar University, International Centre for Theoretical Physics (ICTP), and Aretaieion University Hospital of Athens—for cooperation in a very creative environment.

Authors

Dr. C. S. Sureka is an assistant professor in the Department of Medical Physics at Bharathiar University, Coimbatore, India. She is an associate of the Abdus Salam International Centre for Theoretical Physics (ICTP), Italy. She has been nominated as Radiological Safety Officer (RSO) in Bharathiar University by the Atomic Energy Regulatory Board (AERB), Mumbai, India. During her nine years at Bharathiar University, she has taught various medical physics subjects, including fundamental radiation physics, radiation detectors and instrumentation, medical imaging technology, radiation safety, and radiation biology. She has completed three major research projects and established two laboratories, namely, Radiation Dosimetry and Radiation Protection laboratories for teaching and research. Her research interests are in the areas of Monte Carlo–aided dosimetry for teletherapy and brachytherapy, radiation dosimetry, and biology. She has filed an Indian patent in the area of nanodosimetry. In addition to her professional career, she is a member of the Association of Medical Physicists in India (AMPI), the Asia-Oceania Federation of Organization for Medical Physicists (AFOMP), and the American Association of Physicists in Medicine (AAPM).

Dr. Christina Armpilia is a medical physicist at Aretaieion Hospital, Medical School, National and Kapodistrian University of Athens. During her 13 years in Aretaieion University Hospital, she concerned with three areas of activity: clinical service and consultation, research, and teaching. Her major research projects concern radiobiological modeling with external beam radiotherapy and brachytherapy techniques. Her research interests and scholarly publications are in the areas of clinical radiobiology, medical physics, and radiotherapy. She has taught a variety of medical physics subjects, including courses in radiotherapy, radiobiology, diagnostic radiology, and radiation protection. As author, she has also contributed to a Greek medical oncology book (*Atlas of Thorax Oncology*, 2012, Parisianou ed.). In addition to her professional career, she is a member of the Hellenic Association of Medical Physicists (HAMP) and the European Society for Radiotherapy and Oncology (ESTRO).

1

Cell Biology

Objective

This chapter discusses the basic biological background of the cell that is required to understand the remaining chapters.

1.1 Cell Biology

Cell biology is the study of biochemistry of cells, structure of the cell, cellular components, cell communication with their environment, cell metabolism, their life cycle, cellular abnormalities, etc. In order to understand the biological effects of radiation, it is necessary to review cell biology. Cell theory says that cells are the basic units that perform the functions of living things, so they are called the "building blocks of life." Their major functions include metabolism, growth, irritability, adaptability, repair, reproduction, etc.

An adult human body consists of approximately 100 trillion (100×10^{12}) cells. These cells are different in size, shape, and function. Regarding their size, most of the cells are visible only under the microscope, with dimensions between 1 and 100 μm (10^{-6} m) except human ovum or egg cells. Cells have different shapes like spherical, rectangular, or irregular. Each cell type has its own function in helping our bodies to work properly, and a specific shape and size that helps it to carry out these functions effectively. Our body consists of many specialized cells including bone, muscle, fat, blood, endothelial, exocrine, endocrine, and nerve cells. However, all cells have some similarities in their structure and metabolic activities.

The human body consists of around 200 different kinds of specialized cells. When many identical cells are organized together, it is called a tissue. Tissues are grouped into four basic types, including connective, muscle, nervous, and epithelial tissues. Various tissues organized together for a common purpose form an organ, for example, stomach, skin, brain, uterus, heart, lung, etc. There are almost 78 organs in the human body that are different based on their sizes, shapes, and functions. Two or more organs working together to perform a particular function comprise an organ system. The human body consists of the following 11 organ systems and an individual organism is formed when all these organ systems are grouped together:

1. The circulatory system, which includes heart, veins, arteries, etc.
2. The lymphatic system, which includes nodes, lymph vessels, spleen, etc.

3. The respiratory system, which includes lungs, trachea, etc.
4. The digestive system, which includes stomach, liver, intestines, etc.
5. The endocrine system, which includes pituitary gland, thyroid gland, etc.
6. The nervous system, which includes brain, spinal cord, nerves, etc.
7. The reproductive system, which includes the male testes; the female ovaries, uterus; etc.
8. The skeletal system, which includes bones, cartilage, etc.
9. The muscular system, which includes muscles, tendons, etc.
10. The urinary system, which includes kidneys, bladder, etc.
11. The integumentary system, which includes skin, hair, etc.

1.2 Biochemistry

Biochemistry is the study of life at the molecular level. It is generally classified into three major areas. They are (1) "structural and functional biochemistry," which deals with the discovery of chemical structures, 3D arrangements of biomolecules, and chemicals that are present in living things; (2) "informational biochemistry," which defines the language(s) for storing and transmitting biological data in cells and organisms; and (3) "bioenergetics," which describes the flow of energy in living organisms and how it is converted from one form of energy into another. Since biochemistry is a wide area, a brief review on molecules necessary for life is presented here (Forshier, 2009).

Out of more than 100 chemical elements, only about 31 (28%) occur naturally in plants and animals. Out of 31, 6 of the elements, namely, C, H, O, N, P, and S, are present in major quantities and are very important for life. Twelve of the elements, namely, Na, Mg, Cl, K, Ca, Mn, Fe, Co, Ni, Cu, Zn, and I, are present in small quantities for a specific function. Also, 13 more elements, namely, B, F, Al, Si, V, Cr, Ga, As, Se, Br, Mo, Cd, and W, are present in some organisms. The selection and interaction of these elements during the early stages of evolutionary development of life is not yet unknown. However, it is known that some of these elements are bonded together to produce both inorganic and organic molecules of any organism.

An **inorganic molecule** is generally termed as a substance that does not contain carbon and hydrogen atoms. However, many inorganic compounds like water (H_2O) and hydrochloric acid (HCl) do contain hydrogen atoms. But, only a few inorganic compounds like carbon dioxide (CO_2) contain carbon atoms. Hence, inorganic compounds include **water**, **salts**, **acids**, and **bases**, which are essential in all living organisms. Water (**more than 80%**) plays a tremendous and sensitive role, namely, (1) it dissolves and transports various substances that are necessary for life, (2) it removes waste products from the cells and is hence effectively involved in chemical and metabolic activity, (3) it regulates body temperature, (4) it is involved in the biochemical breakdown of food that we eat, and (5) it serves as an effective lubricant around joints. Salts like NaCl are dissolved in water and dissociate into ions other than H^+ or OH^-. These ions are electrolytes, so they are capable of conducting an electric current in solution. This property is important in transmitting nerve impulses and prompting muscle contraction. Unlike salts, acids release H^+ in solution (quantified in terms of pH), making it more acidic, which is needed for digestion and to kill microbes. But, bases accept H^+, thereby making the solution more alkaline.

An **organic compound** is a substance that contains both carbon and hydrogen. Organic compounds are synthesized by grouping elements *via* covalent bonds. For example, C can form multiple (4) covalent bonds with other C atoms and functional groups such as H, N, O, or S and hence lead to the formation of long and complex chains. This nature of the carbon atom induces the generation of four most important types of organic macromolecules, namely, **nucleic acids (1% of the cell)**, **proteins (15%)**, **carbohydrates (1%)**, and **lipids (2%)**.

Nucleic acids are complex, high-molecular-weight macromolecules made up of thousands of nucleotides (polynucleotide chain). The main function of nucleic acids is to store and transfer genetic information. Each nucleotide is made up of (five carbon) pentose sugars (deoxyribose in DNA and ribose in RNA), phosphate groups, and nitrogenous bases, either purine or pyrimidine. Adenine (A) and guanine (G) are purine bases. Cytosine (C), thymine (T)/uracil (U) are pyrimidine bases. Examples of nucleic acids are **DNA** (deoxyribonucleic acid) and **RNA** (ribonucleic acid). RNA and DNA are discussed later in this chapter.

Proteins are large organic molecules built from a set of amino acids linked by peptide bonds (peptide bonds are formed by the removal of water molecules between two amino acids). Each amino acid consists of CH, a amino group—NH_2, a carboxyl group—COOH, and a variable group—R. Generally, three nucleotides encode an amino acid. Hence, proteins are synthesized by the action of RNAs (mRNA, rRNA, and tRNA) in ribosomes. Using only 22 known amino acid types, a cell constructs thousands of different proteins. The number of amino acids present in a protein ranges from 300 to several thousand. Proteins are called the building blocks of the body based on their seven major functions, namely, (1) they are important in the development and repair of body tissue such as hair, skin, eyes, muscles, and organs; (2) they are a major source of energy; (3) they are involved in the creation of some hormones like insulin (regulates blood sugar), secretin (regulates digestive process), etc.; (4) enzymes are proteins that regulate the rate of chemical reactions in the body; (5) they are involved in the transportation of major molecules (hemoglobin protein for oxygen transport) and storage of molecules (ferritin protein for iron storage in the liver); (6) they form antibodies that help prevent infection, illness, and disease; and (7) they activate DNA. (Remember, all enzymes are proteins except some RNAs but not all proteins are enzymes.)

Carbohydrates are the body's main source of energy. Many of them have the general formula of $(CH_2O)_{n>4}$, and some of them include nitrogen or sulfur in addition to carbon (C), oxygen (O), and hydrogen (H). On the basis of the number of forming units, carbohydrates are classified as monosaccharides, disaccharides, and polysaccharides. Monosaccharides are considered to be simple sugars that cannot be broken down further. Examples are glucose, fructose, and galactose. When two monosaccharides are linked together disaccharides are formed, for example, sucrose (table sugar), lactose (milk sugar), and maltose. Polysaccharides are large macromolecules consisting of many monosaccharide units. Examples include starch and glycogen. Even though carbohydrates are necessary throughout the body, they play a major role in delivering energy to the vital organs, such as the central nervous system, kidneys, brain, liver, heart, etc.

Lipid is the collective name for fats, oils, waxes, and fat-like molecules (such as steroids) found in the body. These molecules consist of carbon, hydrogen, and oxygen that play a variety of roles in the cell: (1) They are involved in energy storage, (2) they are involved in insulation and protection of the body from injury by supporting the immune system, (3) they are essential for proper digestion and absorption of food and nutrients, (4) lipids, namely, phospholipids and cholesterol, are the components of cell membranes, and

(5) cholesterol is a type of lipid required to produce important steroid hormones, which are needed to maintain pregnancy, develop sex characteristics, and regulate calcium levels in the body.

1.3 Structure of the Cell

All the living cells have three basic parts. They are (1) **cell (plasma) membrane**, (2) **cytoplasm** with cell organelles, and (3) **nucleus** with DNA. The cytoplasm and the nucleus of a cell collectively termed as "protoplasm." Generally, cells are divided into two types. They are prokaryotic cells and eukaryotic cells. Prokaryotic cells do not have membrane-bound organelles. They have DNA, but they are not bound inside the nucleus, for example, bacterial cells. In contrast, eukaryotic cells are larger and more complex cells than prokaryotic cells. They have a true nucleus containing DNA as well as various other membrane-bound organelles, for example, plant and animal cells.

Plant and animal cells have many similarities and differences. The similarities are the presence of the cell membrane and cell organelles. The differences are that animal cells do not have a cell wall or chloroplasts like plant cells and that animal cells are round and irregular in shape, but plant cells have fixed, rectangular shapes. Even though human cells are varied in its shape, size, and function, they all possess these three basic parts of the cell except the mature red blood cells (do not have nucleus). The crosssectional view of an animal cell (human cell) along with its cellular components is depicted in Figure 1.1 and its components are discussed here.

1.4 Cellular Components

1.4.1 Cell Membrane or Plasma Membrane

The cell membrane is a thin, flexible, semi-permeable membrane. It is a barrier that separates cytoplasm from its exterior surroundings. The fluid mosaic model describes the structure of a cell membrane (a part of Figure 1.1). It is composed of a phospholipid bilayer, proteins, some cholesterols, and carbohydrates. A single phospholipid molecule has two different ends: a head and a tail. The head end contains a phosphate group and is hydrophilic (likes water molecules). The tail end is made up of two strings of hydrogen and carbon atoms called fatty acid chains and these chains are hydrophobic (do not like water molecules). This nature of phospholipid bilayer permits only lipid soluble nutrients and wastes into/out of the cell.

Some proteins are located inside of the lipid bilayer (integral proteins) and some outside of the lipid bilayer (peripheral proteins). These proteins monitor and correct the cell's chemical composition and also communicate with the external environment through the use of hormones, neurotransmitters, pores, and other signaling molecules. Cholesterol molecules are hydrophobic so they are attached to the hydrophobic tails in the lipid bilayer. These molecules keep the phospholipid tails from coming into contact and maintain the oil-like and flexible nature of the membrane. Carbohydrates, or sugars, are sometimes

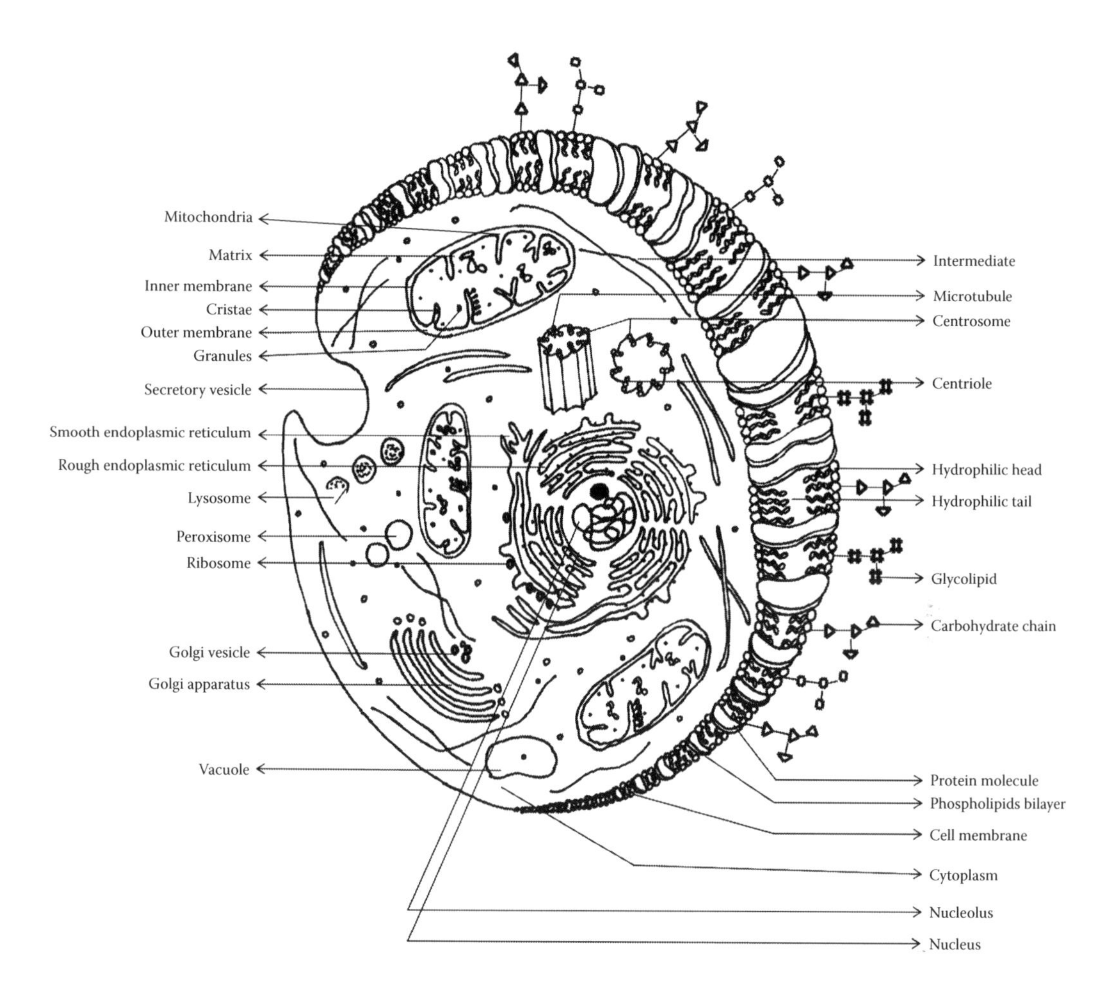

FIGURE 1.1
The crosssectional view of the animal cell structure along with its cellular components.

attached to proteins or lipids on the extracellular side of a cell membrane. It provides protection to the plasma membrane and is also important in cell recognition.

It also consists of many structures, such as caveolae (flask-shaped structures), ion channels, and membrane pumps. The main functions of the cell membrane are (1) to maintain the physical integrity of the cell, and (2) to control the movement of particles (ions or molecules) by diffusion through ion channels, active transport through membrane pumps, endocytosis (into the cell), and exocytosis (out of the cell) through caveolae.

1.4.2 Cytoplasm

The cytoplasm represents everything inside the cell except the nucleus. Both organic and inorganic molecules are present in the cytoplasm. Major processes necessary for life, such as (1) protein synthesis, (2) cellular respiration, and (3) metabolism (Metabolism = Catabolism + Anabolism) take place in the cytoplasm.

The cytoplasm has three major elements, namely, (1) cytosol, (2) cytoplasmic inclusions, and (3) organelles. Cytosol is a clear, thick, jellylike material consisting of water, salts, and organic molecules. Cytoplasmic inclusions are small particles of insoluble substances, such as minerals, microtubules (the cell's cytoskeleton), and lipid droplets (composed of lipids and proteins) present in the cytosol. Organelles (small organs) are sub-structures of the cell, which include mitochondria, endoplasmic reticulum, ribosomes, Golgi apparatus, lysosomes and peroxisomes, vacuoles, and centrosomes.

1.4.2.1 Mitochondria

Mitochondria are well-defined, double-membrane, oval or rod-shaped cytoplasmic organelles of the cell. They are called the "power houses" of the cell. Mitochondria have an outer membrane and a highly folded inner membrane (crista) (a part of Figure 1.1). The double membrane of the mitochondria contains enzymes, which assist in breaking down carbohydrates, fat, and proteins into energy, and are stored in the cell as ATP (adenosine-triphosphate). The shape, size, and number of mitochondria vary from one cell to another depending upon their needs. Cells in the heart, pancreas, kidney, skeletal muscles, etc., require more energy to perform their functions so that they contain a large number of mitochondria.

1.4.2.2 Endoplasmic Reticulum

The endoplasmic reticulum (ER) is an interconnected network of tubules, vesicles (cellular chemical reaction chambers for organizing cellular substances), and cisternae (flattened sacs) within cells. It extends from the nuclear membrane to the cell membrane. The functions of the endoplasmic reticulum are synthesis and export of proteins and membrane lipids. There are two types of ER. They are rough ET and smooth ET. Rough ER consists of a lot of ribosomes, which give a rough appearance. But, ribosomes are not present in smooth ER (a part of Figure 1.1). The main role of the rough ER is the synthesis of proteins (enzymes). But, the smooth ER synthesizes lipids and steroids, metabolizes carbohydrates, and regulates the calcium concentration and drug metabolism (drug inactivation too). The quantity of both rough and smooth ER in a cell depends upon the metabolic activities of the cell.

1.4.2.3 Ribosomes

Ribosomes are the **protein synthesizers** of the cell. A ribosome is made from complexes of RNAs and proteins, and hence it is referred to as ribonucleoprotein. Ribosomes consist of two major components—the small ribosomal subunit, which reads the RNA, and the large subunit, which joins amino acids to form a polypeptide chain (protein). Ribosomes are found in (1) the cytosol in the cytoplasm, (2) the endoplasmic reticulum to make proteins that will be used inside of the cell, and (3) the nuclear membrane to synthesize proteins that are released into the perinuclear space.

1.4.2.4 Golgi Apparatus/Golgi Bodies/Golgi Complex

The Golgi apparatus is a series of flattened disk-like structures (or cisternae) located close to the endoplasmic reticulum. The Golgi apparatus receives proteins and lipids (fats) from the endoplasmic reticulum. Then, it modifies some of those proteins and lipids, and sorts, concentrates, and packs them into sealed droplets called Golgi vesicles. Depending on the contents, these vesicles are transferred to (1) the cytoplasm of the cell for lysosome creation, (2) the plasma membrane of the cell, and (3) outside the cell for secretion. The number of Golgi apparatuses varies in the cells, and it is rich in gastric gland cells, salivary glands, and pancreatic glands.

1.4.2.5 Lysosomes and Peroxisomes

Lysosomes are single-membrane organelles containing digestive enzymes. They digest excess or worn-out organelles, food particles, dead cells, foreign materials, viruses, and bacteria. Initially, enzyme proteins are created in the rough endoplasmic reticulum and then sent to the Golgi apparatus for packing. The Golgi apparatus then creates the digestive enzymes and pinches off small, very specific vesicles. These vesicles are called lysosomes. Once the lysosomes are created by the Golgi apparatus, they float in the cytoplasm until their requirement.

Peroxisomes contain at least 50 enzymes and are separated from the cytoplasm by a single, small, and rounded lipid bilayer membrane. They are called peroxisomes because they all produce hydrogen peroxide by breaking organic molecules. A peroxisome protein is involved in preventing the occurrence of kidney stones. If these lysosomes and peroxisomes burst, they begin to digest the cell's protein, causing cell death. For this reason, they are also called as "**suicide bags**."

1.4.2.6 Vacuoles

Vacuoles are **storage bubbles** present in all cells. Vacuoles store food or nutrients necessary for a cell to survive. They can also store waste products in order to protect other parts of cells from contamination. Later, those waste products would be sent out of the cell.

1.4.2.7 Centrosomes

Centrosomes (shown in the Figure 1.1) are composed of two centrioles (cylindrical structures located near the nucleus) and can produce microtubules (key component of the cytoskeleton) of a cell. During cell division, the pair of centrioles detach from each other. During this process, thin cytoplasmic spindle fibers are formed between the centrioles. These spindle fibers are connected to the specific chromosome in order to distribute the number of chromosomes between two daughter cells equally.

1.4.3 Nucleus

Nucleus is an information center of the cell. It is a membrane-bound structure that controls the cell's metabolic activity and cell division by regulating gene expression. It is spherical in shape, floats in the cytoplasm, and is generally located in or near the center of the cell. It consists of (1) a nuclear membrane with pores, (2) nucleoplasm, (3) subnuclear organelles, and (4) the most important DNA/chromosomes.

The **nuclear membrane** is the phospholipid bilayer membrane located around the entire contents of the nucleus and is connected with the rough endoplasmic reticulum. It consists of thousands of nuclear pores. These pores regulate the movement of macromolecules like proteins and RNA from nucleus to protoplasm and vice versa. However, it freely permits the passage of water, ions, ATP, and other small molecules. In this way, the nuclear membrane helps to maintain the shape of the nucleus. It also has some control over the information flow into the cell since cellular information is carried by the macromolecules.

The **nucleoplasm** is a highly viscous fluid present inside the nuclear membrane. It acts as a suspension fluid for the chromosomes and other subnuclear organelles. It is made up of mostly water, a mixture of various biomolecules such as nucleotides (building blocks of nucleic acids) and enzymes, and also dissolved ions. A network of fibers known as the nuclear matrix is also present in the nucleoplasm. It helps to maintain the shape and structure of the nucleus. It is also responsible for the transport of biomolecules that are important to cellular function.

The **subnuclear organelles** are specialized regions or nonmembranous small bodies having proteins and RNA. It includes nucleoli or nucleolus, Cajal bodies, and speckles. The nucleolus is the major site for the synthesis of ribosomal subunits and assembles ribosomes. Cajal bodies are found in the nucleus of proliferative cells like embryonic cells and tumor cells or metabolically active cells. Speckles are enriched in pre-messenger RNA splicing factors and are located in the interchromatin regions of the nucleoplasm.

The **DNA** (deoxyribo nucleic acid) is a very long (can be extended up to 6 m), thread-like, and spiral-shaped biomolecule having a double helical structure. It is richly present in the cell nucleus and makes the chromosomes in association with proteins. However, it is also present in the mitochondria, plasma membrane, and centrioles. It consists of chains of nucleotide units. As discussed earlier, each nucleotide unit contains three components: the deoxyribose sugar, a phosphate group, and a nitrogen-containing base, either purine or pyrimidine. Adenine (A) and guanine (G) are purine bases. Cytosine (C) and thymine (T) are pyrimidine bases. In DNA, A always pairs with T through 2H bonds, and G always pairs with C through 3H bonds as shown in Figure 1.2 (to remember, **ATGC**—**A**ll **T**raditional **G**enetic **C**ode). DNA has four major functions, which are as follows: (1) It contains the blueprint for producing proteins/ enzymes, (2) it regulates the synthesis of proteins, (3) it carries the genetic information during cell division, and (4) it transmits that information from parental organisms to their offspring. Even though DNA controls the synthesis of proteins, its function is controlled by proteins too. All the functions of DNA are supported by RNA.

RNA is synthesized in the nucleus taking DNA as a template by the action of an enzyme RNA polymerase (protein). This process is known as transcription. RNA is found in the nucleoli, cytoplasm, and some cell ribosomes. Unlike DNA, RNA consists of a single-stranded, shorter nucleotide chain, which contains ribose sugar (deoxyribose in DNA), and the complementary base to adenine is uracil (thymine in DNA). There are three subunits of RNA. They are messenger RNA (m-RNA), transfer RNA (t-RNA), and ribosomal RNA (r-RNA). Messenger RNA (mRNA) is the RNA that carries information from DNA to the ribosome (in the protoplasm) for protein synthesis (known as translation) in the cell. Transfer RNA (tRNA) is a small RNA chain of about 80 nucleotides that transfers a specific amino acid to a growing polypeptide chain at the ribosomal site of protein synthesis during translation. Ribosomal RNA (rRNA) assists in the linking of the messenger RNA to the ribosome.

As shown in Figure 1.3, approximately 147 base pairs of DNA molecules are wrapped around the disk-shaped core of proteins (histones) forming nucleosomes. Large numbers of nucleosomes are condensed together to form chromatins (resting/non-dividing time). During the process of cell division, these chromatins undergo further condensation and appear as short, rod-like structures in the nucleus called "**chromosomes**." Hence, chromatins are lower-order nucleosome sequences and chromosomes are higher-order nucleosome sequences. Chromosomes (not chromatins) are visible during the process of cell division as two identical sister chromatids (arms) attached with centromeres. So each chromosome consists of a single molecule of DNA associated with an equal mass of proteins.

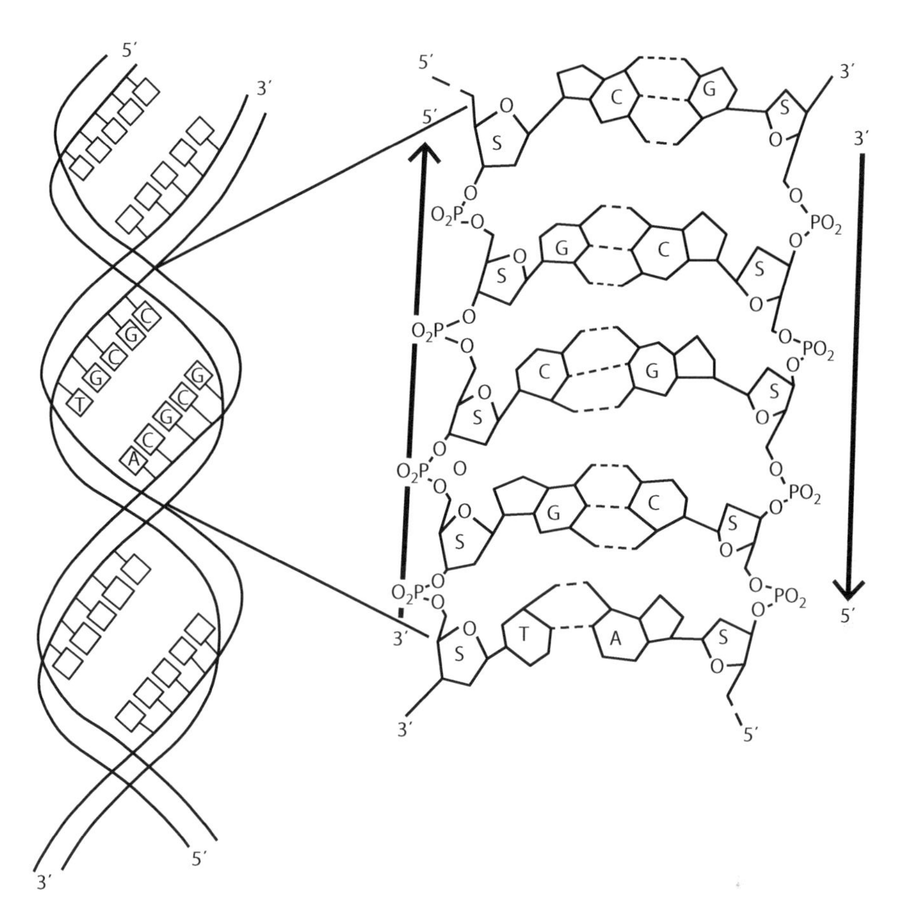

FIGURE 1.2
Structure of the DNA.

The chromosomes carry genes. A gene is a specific sequence of base pairs (segment of DNA) that contains the genetic information for a specific function. The position of a gene in the chromosome is referred to as a "locus". The human genome contains approximately 6×10^9 base pairs of DNA in 23 pairs of chromosomes within the nucleus of all somatic human cells. The total number of protein-encoding genes is in the range of 25,000–50,000 per haploid set of chromosomes (Pertea and Salzberg, 2010).

In plants and animals, there are two major types of cells: germ cells and somatic cells. Germ cells are involved in reproduction, for example, egg cells in female and sperm cells in male. All other cells in the body are somatic cells. Each human somatic cell contains two complete sets of chromosomes (one from each parent). This number is known as the diploid (2n) number. For example, in humans, somatic cells contain 46 chromosomes organized into 23 pairs. But germ cells have unpaired chromosomes and are known as the haploid (n) number. In humans, this number is 23 unpaired chromosomes. Out of 23 pairs of chromosomes in human cells, there are two types: 22 pairs are autosomes and 1 pair is allosome (sex chromosome). Sex chromosomes in females are homozygous (XX) and in males are heterozygous (XY).

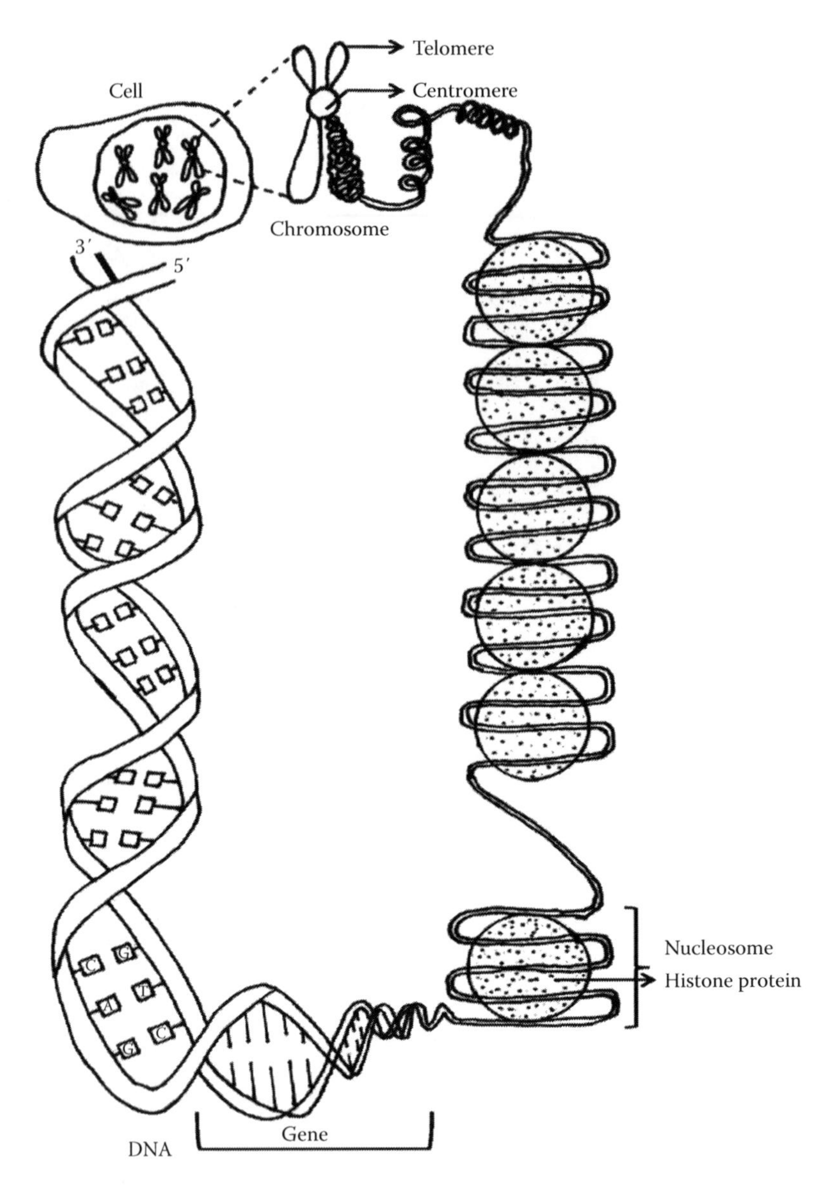

FIGURE 1.3
Formation of chromosomes from DNA.

1.5 Communication of Cells with Their Environment

Cells do not exist alone. They communicated with their environment to form tissues and perform activities, such as cell migration, cell growth, cell differentiation, 3D organization of tissues and organs, etc. This cellular environment (Figure 1.4) consists of (1) **glycocalyx,** (2) **extra cellular matrix** (ECM), and (3) **cell adhesion molecules** (CAM) (Marthy, 2013).

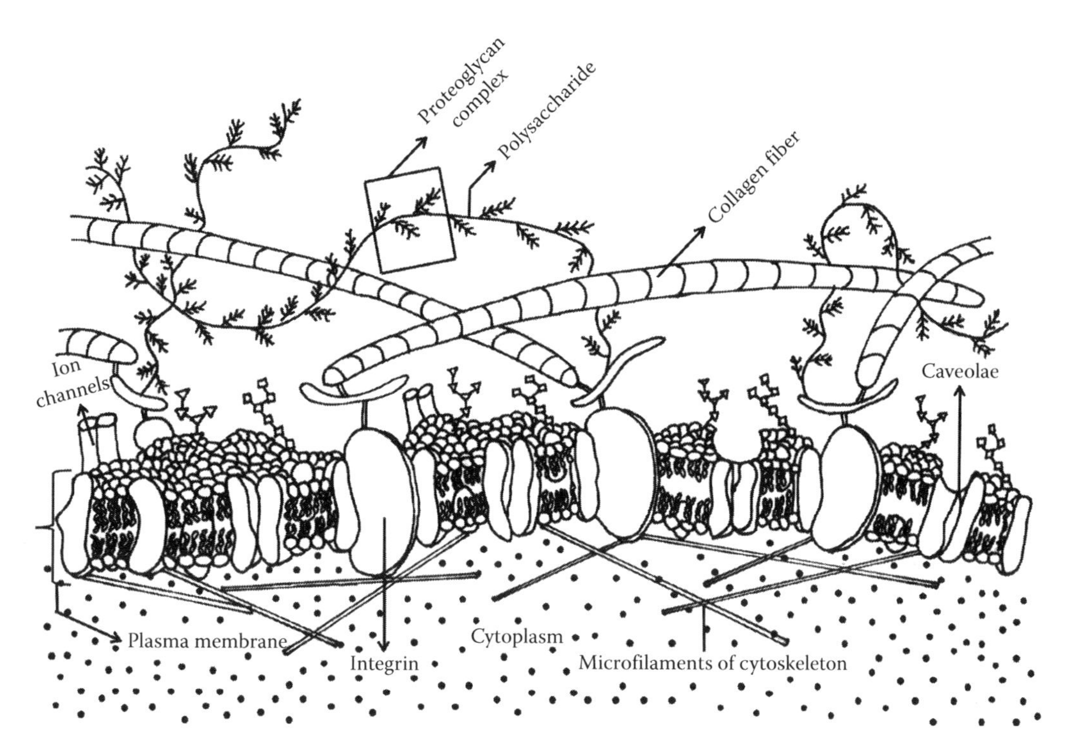

FIGURE 1.4
Cellular environment.

The **glycocalyx** (sugar coat) is formed from carbohydrate projections from the plasma membrane. It (1) functions as a mediator for cell-to-cell and cell-to-substrate interactions, (2) functions as a barrier to particles, which move toward the plasma membrane, (3) provides mechanical protection to cells, and (4) binds important regulatory factors.

The extracellular matrix (ECM—the substance between cells) is a gel-like "ground substance" present around the cell. The ECM is composed of structural proteins (collagen and elastic), proteinpolysaccharide complexes, and cell adhesive glycoprotein molecules. The ECM surrounds the cell as fibrils (that contact the cells on all sides) or as a sheet called the basement membrane. The ECM provides mechanical support and acts as a biochemical barrier and a medium for (1) extracellular communication, (2) stable positioning of cells in tissues, and (3) repositioning of cells by cell migration during cell development and wound repair. Depending upon the requirement, these ECMs can be calcified to form bones and teeth as they contain structural proteins.

The cell adhesion molecule (CAM), which is present in ECM, belongs mainly to a family of chemicals called glycoproteins. They are located at the cell surface and form different types of complexes and junctions to join one cell with another, and join cells to the ECM and the ECM to the cell cytoskeleton. (The cytoskeleton is an intracellular matrix that supports cell shape and function.) The various types of junctions are (1) **tight junctions**—they pull the walls of two cells very close together but do not allow molecules to pass from one cell to another, and (2) **gap junctions**—these join two cells together with a cluster of fine tubes and allow small molecules to pass from one cell to another. Hence, CAMs assist in

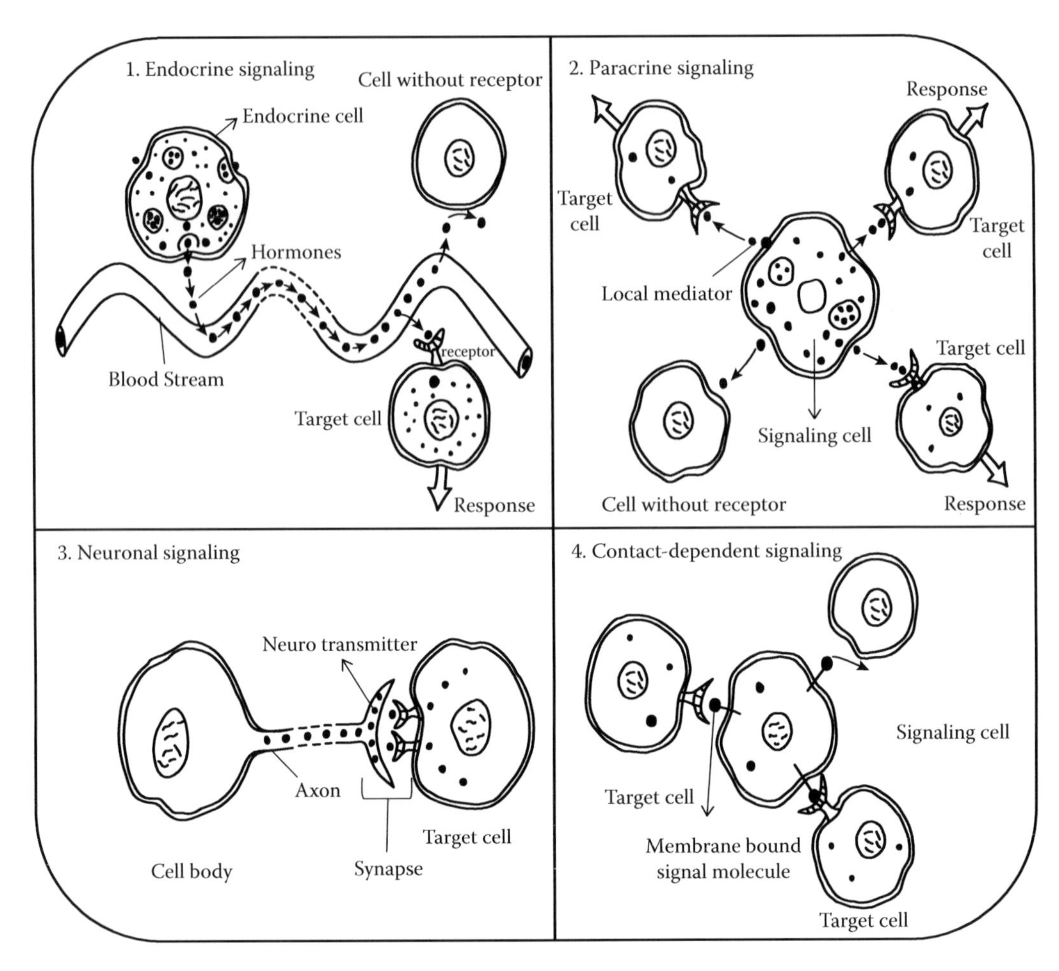

FIGURE 1.5
Cell signaling.

(1) the adhesion of cells to one another to provide organized tissue structure, (2) the transmission of extracellular signals across the cell membrane, and (3) the migration of cells through the regulation of a cell-adhesive molecular composition. The increase or decrease in the ECM and CAM leads to many disorders and diseases, such as common colds, HIV, malaria, **cancer**, asthma, inflammatory diseases, viral infections, etc.

Due to their nature, cells communicate with their environment in several ways. The four important ways of cell signaling are (1) endocrine signaling (cells at a longer distance)—hormones are released into the circulatory system, which carries them to the target cell that contains receptors to receive signals; (2) paracrine signaling (nearby cells)—secretions from one cell have an effect only on cells in the immediate area; (3) neuronal signaling (cells at longer distance)—involves the transmission of signal molecules, called neurotransmitters, from a neuron over a small synaptic gap to the target cell; and (4) direct contact (cells in contact)—two cells in direct contact with each other can send signals across gap junctions (Figure 1.5).

1.6 Cell Metabolism

The metabolism represents a network of highly coordinated enzyme-catalyzed chemical reactions within the cells of living organisms. Most reactions catalyzed by enzymes can be classified into the following four categories (Smith and Wood, 1991). These reactions allow organisms to grow and reproduce, maintain their structures, and respond to their environments.

1. **Oxidation–reduction reactions**: These are the most important reactions in energy extraction and transfer in cells. These reactions transfer electrons from one molecule to another. A molecule that gains electrons is said to be **reduced**. One way to remember this is that adding negatively charged electrons reduces the electric charge on the molecule. Conversely, molecules that lose electrons are said to be **oxidized**. (To remember, use the mnemonic **OIL RIG**—**O**xidation **I**s **L**oss [of electrons], **R**eduction **I**s **G**ain.)
2. **Hydrolysis–dehydration reaction**: These reactions are important in the breakdown and synthesis of large biomolecules. In a **hydrolysis reaction** (hydro- water + lysis- to loosen or dissolve), a substrate is changed into one or more products through the addition of water. In these reactions, the covalent bonds of the water molecule are broken ("lysed") so that the water reacts as a hydroxyl group OH^- and a hydrogen ion H^+. For example, an amino acid can be removed from the end of a peptide chain through a hydrolysis reaction (when an enzyme name consists of the substrate name plus the suffix -ase, the enzyme causes a hydrolysis reaction). In **dehydration reactions** {de-, out + hydr-, water}, a water molecule is one of the products. In many dehydration reactions, two molecules combine into one, losing water in the process. In the process, one substrate molecule loses a hydroxyl group –OH and the other substrate molecule loses a –H to create water, H_2O. When a dehydration reaction results in the synthesis of a new molecule, the process is known as dehydration synthesis.
3. **Addition–subtraction and exchange reaction**: An **addition reaction** adds a functional group to one or more of the substrates. A subtraction reaction removes a functional group from one or more of the substrates. Functional groups are exchanged between or among substrates during exchange reactions. For example, removal of an amino group from an amino acid or peptide is a deamination reaction. Addition of an amino group is amination, and the transfer of an amino group from one molecule to another is transamination.
4. **Ligation reactions**: Ligation reactions join two molecules together using enzymes known as synthetases and energy from ATP. It is an essential laboratory procedure in the molecular cloning of DNA whereby DNA fragments are joined together to create recombinant DNA molecules.

Metabolism is generally divided into two basic processes. They are (1) catabolism and (2) anabolism (Figure 1.6). **Catabolism** (destructive metabolism) is the process of breaking up large molecules (mostly carbohydrates and fats) into more simple molecules and produces energy by the way of cellular respiration (respiration is the process of oxidizing food molecules, like glucose, to carbon dioxide and water) and heat. This energy is used as fuel

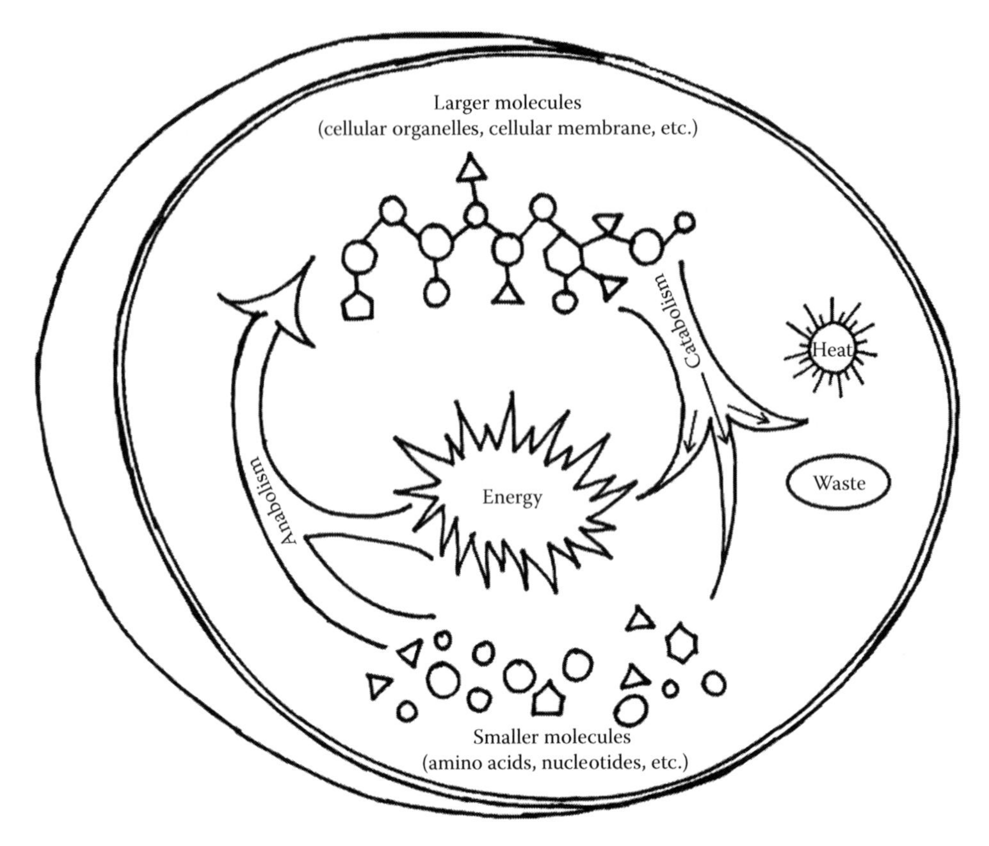

FIGURE 1.6
Cellular metabolism.

for anabolism, heats the body, and enables the muscles to contract and the body to move. It also produces waste products, such as CO_2 etc., which is removed from the body through the skin, kidneys, lungs, and intestines. **Anabolism** (constructive metabolism) is the process of building and storing large biomolecules (proteins and nucleic acids) from small molecules (amino acids and nucleotides) using the energy generated from catabolism. Hence, it supports the growth of new cells, the maintenance of body tissues, and the storage of energy for future use. Anabolic and catabolic reactions take place simultaneously in cells throughout the body so that at any given moment, some biomolecules are being synthesized while others are being broken down. The energy released from or stored in the chemical bonds of biomolecules during metabolism is commonly measured in kilocalories (kcal).

Cells regulate their metabolic pathways in the following five basic ways:

1. By controlling enzyme concentrations. Factors that affect the activity of the enzyme include (a) the temperature, (b) the pH of an enzyme, (c) the concentration/amount/volume of the enzyme, and (d) the concentration/amount/volume of the substrate. When these factors are increased, the reaction rate is also increased.
2. By producing modulators that change reaction rates.

3. By using two different enzymes to catalyze reversible reactions. Cells can use reversible reactions to regulate the rate and direction of metabolism.
4. By compartmentalizing enzymes within intracellular organelles. Many enzymes of metabolism are only present in specific subcellular compartments. For example, enzymes of carbohydrate metabolism are dissolved only in the cytosol. This allows the cell to control metabolism by regulating the movement of the substrate from one cellular compartment to another.
5. By maintaining an optimum ratio of ATP to ADP to regulate the energy status.

1.7 Life Cycle of the Cell

1.7.1 Cell Cycle

The millions of cells in our body are formed from a single cell through many sequential cyclical patterns of cell division. The rate of cell division is varied with respect to the type of cell and organism as well. For example, stomach-lining cells take 2–7 days, skin epidermis cells take 10–30 days, and red blood cells take about 4 months (so donate blood once in 4 months only) to renew. But, nerve cells, muscle cells (heart), and eye lens cells do not divide at all. This sequential cyclical pattern of events by which a cell duplicates its genome, synthesizes other constituents of the cell, and eventually divides into two daughter cells is referred to as cell cycle.

Initially, the cell cycle (Figure 1.7) is divided into two major phases, namely, (1) interphase, and (2) M phase (mitosis or meiosis). During the interphase, the cell prepares itself for division by undergoing both cell growth and DNA replication in an orderly manner. The interphase is further divided into three phases. They are (1) the G_1 phase (Gap 1), (2) the S phase (DNA synthesis phase), and (3) the G_2 phase (Gap 2).

The G_1 phase corresponds to the interval between the M phase and the S phase. During the G_1 phase, the cell is metabolically active and continuously grows by synthesizing components (like proteins) required for DNA synthesis.

During the S or the synthesis phase, DNA synthesis/replication takes place in the nucleus and the centriole also duplicates in the cytoplasm. So, the amount of DNA per cell doubles during the S phase, that is, if the initial amount of DNA is denoted as 2C then it increases to 4C. However, there is no increase in the chromosome number; if the cell had diploid or 2n numbers of chromosomes at G_1, even after the S phase the number of chromosomes remains the same, that is, 2n. But, after the S-phase of the cell cycle, each chromosome consists of two chromatids and the chromatids are connected by a centromere. Within a chromosome the two chromatids bear identical DNA base sequences as each is made by DNA replication. Upon completion of cell division, each chromatid becomes a separate chromosome. During the G_2 phase, proteins are synthesized in order to prepare the cell for cell division, so the cell grows continuously.

Generally, the time taken by fast-dividing mammalian cells to complete a single cell cycle is approximately 24 hours. (Generally, it ranges between 10 and 40 hours depending upon the type of fast dividing cell.) It includes about 8–10 hours for the G_1 phase (long), 5–6 hours for the S phase, 6–8 hours for the G_2 phase (in total 18–20 hours for the interphase), and about 2 hours for cell division in the M phase.

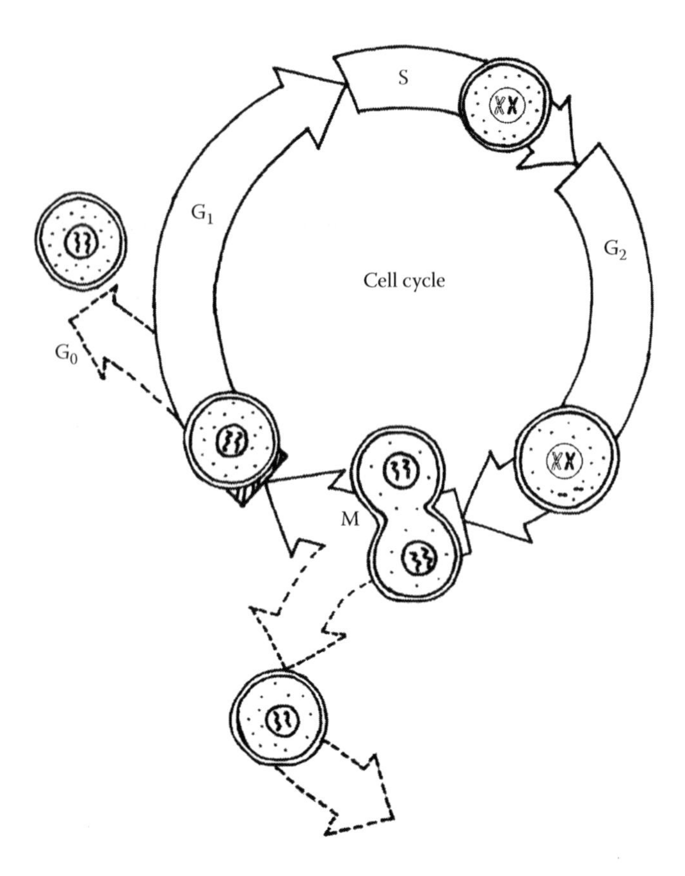

FIGURE 1.7
Cell cycle.

Some cells do not undergo cell division (e.g., heart cells) since they lack centrioles, and many other cells divide only occasionally. These cells further stay in the G_1 phase and then enter an inactive stage called the quiescent stage (G_0) of the cell cycle. During this G_0 phase, the cell cycle is arrested. But the cells in this stage remain metabolically active but no longer proliferate unless and otherwise they are triggered depending on the requirement of the organism.

Then, the active cells enter into the last and most important phase of the cell cycle, that is, the M phase. The M phase starts with the nuclear division, corresponding to the separation of daughter chromosomes (termed as karyokinesis) and usually ends with the division of cytoplasm (cytokinesis), in turn forming two separate cells. Depending upon the type of cell, it undergoes either mitotic or meiotic cell division. For examples, somatic cells undergo mitotic cell division but germ cells undergo meiotic cell division.

1.7.2 Cell Division

1.7.2.1 Mitosis

Mitotic cell division occurs in somatic cells of both sexually and asexually reproducing organisms. The cell divides only once and the chromosomes also divide only once. The chromosome number remains constant at the end of mitosis and there is no crossing over.

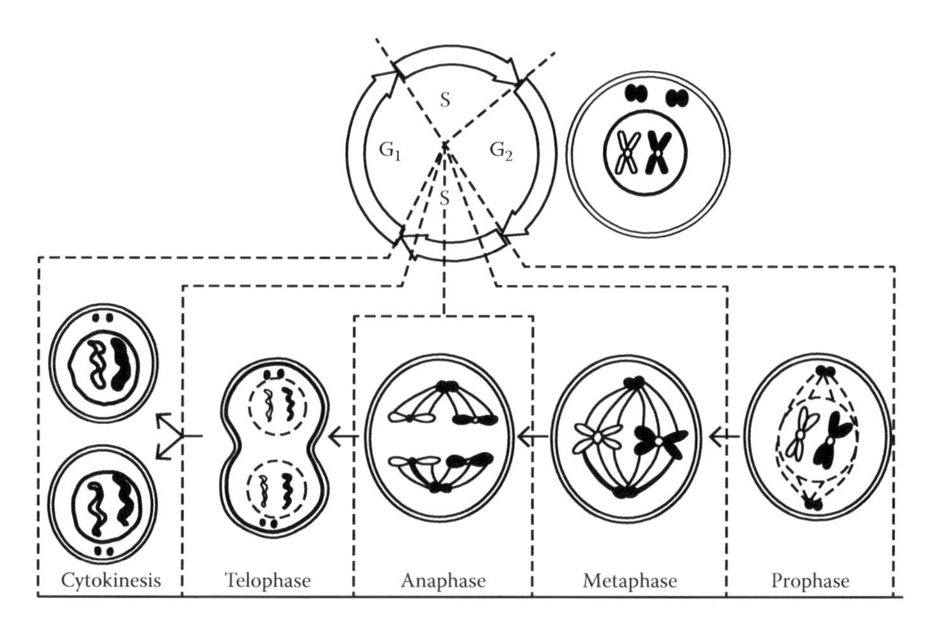

FIGURE 1.8
Mitotic cell division.

It helps in the multiplication of cells and also takes part in healing and repair. Mitosis is divided into the following four stages (Figure 1.8): (1) **prophase** (2) **metaphase** (3) **anaphase**, and (4) **telophase**.

Prophase is the first stage of mitosis following the G_2 phase of interphase. In this phase, chromosomal materials are condensed, and it is visible under the microscope. Then, the centriole, which has duplicated during the S phase of cell cycle, now begins to move toward the opposite poles of the cell. Hence, at the end of the prophase, the nuclear membrane begins to disappear, and the nucleolus is no longer visible.

During the **metaphase** the paired chromosomes are lined up at the equator of the cell where the centromeres orient toward the equator while chromatids orient toward the poles. Spindle fibers from each centriole attach to the centromeres of the chromosomes. The nuclear membrane disappears entirely.

During the **anaphase**, centromeres are divided and hence the two sister chromatids are separated. This is referred to as chromosome segregation or karyokinesis. Then, these chromatids start to move toward the opposite poles.

During the **telophase**, chromosome clusters are formed at opposite spindle poles and they appeared to be much longer, thinner, and indistinct elements (decondensed). Spindle fibers disappear. Nuclear envelope assembles around the chromosome clusters. The nucleolus, Golgi complex, and ER are also reformed.

It is followed by the division of cytoplasm, which is referred to as **cytokinesis**. Hence, at the end of mitosis, the parental cell is divided into two daughter cells with the same genetic information.

1.7.2.2 Meiosis

Meiotic cell division takes place in the germ cells of sexually reproducing organisms. It is a unique type of cell division that takes part in the formation of germ cells (megaspores–haploid spores) in the gonads (ovaries and testes) of sexually matured people during gametogenesis. But, it is not involved in cell multiplication. During meiosis, cell division occurs twice, but the DNA replication occurs only once. So the chromosome number is reduced from the diploid (2n = 46) to the haploid (n = 23). Crossing over or exchanges of similar segments between nonsister chromatids of homologous chromosomes also occur.

Meiotic cell division is classified into two major sequences. They are (1) Meiosis I and (2) Meiosis II. Both Meiosis I and Meiosis II consist of four major phases (Figure 1.9):

Prophase I: Homologous chromosomes further condense and pair. Chromosome crossing over occurs. Spindle fibers form between centrioles, which move toward opposite poles.

Metaphase I: The bivalent chromosomes align along spindle equator where centromeres orient toward poles while chromatids orient toward the equator. The microtubules from the opposite poles of the spindle attach to the pair of homologous chromosomes.

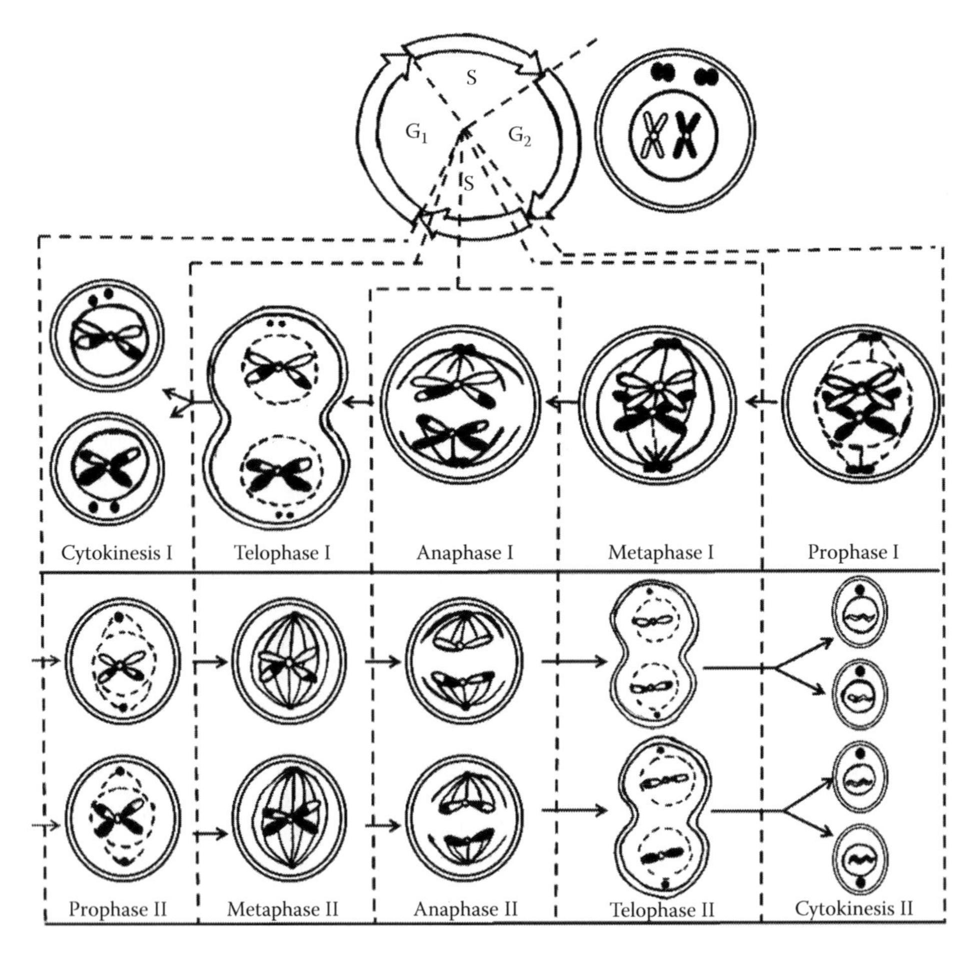

FIGURE 1.9
Meiotic cell division.

Anaphase I: Homologous pairs of chromosomes separate and move to opposite poles. But, the sister chromatids remain associated at their centromeres.

Telophase I: The nuclear membrane and nucleolus reappear, and cytokinesis follows. Each cell receives exchanged chromosomal material from homologous chromosomes.

Prophase II: Chromosomes condense again. Spindle fibers form between centrioles that move toward opposite poles.

Metaphase II: Microtubule spindle apparatus attaches to chromosomes. Chromosomes align at the equator along the spindle.

Anaphase II: It begins with the simultaneous splitting of the centromere of each chromosome (which was holding the sister chromatids together), allowing them to move toward the opposite poles of the cell.

Telophase II: Meiosis ends with telophase II in which the two groups of chromosomes once again get enclosed by a nuclear envelope and cytokinesis results in the formation of four haploid (half the original number of chromosomes) granddaughter cells.

1.7.3 Cell Synchronization

Cell synchronization is a process by which cells at different stages of the cell cycle are brought to the same phase. Thus, a synchronized culture is one in which cells of similar age progress as a group through the division.

Under normal circumstances, progression through the cell cycle will be asynchronous in the tumor mass. However, a tumor is a complex dynamic object that can adapt to changes in its environmental conditions, such as radiotherapy. This change can induce a synchronization of the cell cycle and consequently increase the efficiency of the radiotherapy due to induced radiosensitivity in tumors. Such kind of synchronization plays a vital role in fractionated radiation therapy. But, it is still at an infant stage to implement it clinically.

Cell synchronization is also required to study the progression of cells through the cell cycle in a culture. There are two well-described methods to synchronize the cells. They are (1) the physical method and (2) the chemical method (Lehr, 2002).

The physical method is based on the cell density, cell size, sedimentation velocity, affinity of antibodies on the cell surface, and light scattering or fluorescence emission by labeled cells. The physical characteristics, such as cell size and sedimentation velocity, are operative in the technique of centrifugal elutriation. Fluorescence activated cell sorting (FACS) is a technique for sorting out the cells based on the differences that can be detected by light scattering (e.g., cell size) or fluorescence emission (by penetrated DNA, RNA, proteins, and antigens).

In the chemical method, the cells can be separated by blocking metabolic reactions. Two types of metabolic blockades are in use. They are (1) inhibition of DNA synthesis and (2) nutritional deprivation. In the method of inhibition of DNA synthesis, DNA synthesis can be inhibited/suppressed by chemical inhibitors such as thymidine, aminopterin, hydroxyurea, and cytosine arabinoside during the S phase of cell cycle. Hence, the cell cycle is predominantly blocked in the S phase that results in viable cells. In the method of nutritional deprivation, the serum from the culture medium is eliminated for about 24 hours, which results in the accumulation of cells at the G_1 phase. This is termed as the "G_1 block."

1.8 Cellular Abnormalities

Due to the variation in the composition of biomolecules present in the cells and also due to other external stimuli, the characteristics of the normal cells, such as (1) reproducing themselves only when and where they are needed, (2) sticking together in the right place in the body, (3) self-destructing if they are damaged or too old, and (4) becoming specialized (mature), are modified abnormally. This leads to many diseases, including cancer. Since medical physics deals with the use of ionizing radiation for diagnosis and treatment of disease, especially cancer, importance is given toward cancer in Chapter 2.

Objective-Type Questions

1. The major function of a cell is
   a. Metabolism
   b. Irritability
   c. Adaptability
   d. All of the above
2. __________ is an inorganic molecule
   a. Water
   b. Protein
   c. Lipid
   d. Carbohydrate
3. In the human body, nucleic acids are present in the following percentage.
   a. 1%
   b. 5%
   c. 10%
   d. 15%
4. The protein synthesizers of the cell are
   a. Golgi apparatus
   b. Lysosomes
   c. Ribosomes
   d. Mitochondria
5. How many base pairs constitute a nucleosome?
   a. 142
   b. 147
   c. 152
   d. 157

6. Cell signaling initiated by releasing hormones into the circulatory system is referred to as
   a. Endocrine signaling
   b. Paracrine signaling
   c. Neuronal signaling
   d. Direct contact
7. A substrate is changed into one or more products through the addition of water is referred to as
   a. Oxidation
   b. Reduction
   c. Hydrolysis
   d. Dehydration
8. ________ is the longest phase in the cell cycle.
   a. G_1 phase
   b. S phase
   c. G_2 phase
   d. M phase
9. Chromosome crossing over takes place during the following phase of the cell cycle.
   a. Prophase I
   b. Metaphase I
   c. Anaphase I
   d. Telophase I
10. During which phase of the cell cycle can DNA synthesis be inhibited/suppressed by using chemical inhibitors?
    a. G_1 phase
    b. S phase
    c. G_2 phase
    d. M phase

2

A Brief Review of Cancer

Objective

Since the field of medical physics mainly deals with cancer diagnosis and treatment using radiation, a brief review of cancer is provided in order to understand the effect of radiation on cancer cells and tissues.

2.1 Definition

Cancer (the uncontrolled growth and spread of unwanted cells) is a disease in which abnormal cells grow in an uncontrolled manner and spread to a nearby region. It is an infectious disease, a disease of aging, an environmental disease, a genetic disease, and it is a collection of more than 100 types of diseases.

Normally, our body forms new cells as it needs them, and replacing old cells that die. Sometimes this process goes wrong. New cells grow even when we do not need them, and old cells do not die when they should. These extra cells can form a mass called a tumor. Tumors can be benign or malignant. Benign tumors are not cancers while malignant ones are. Cells from malignant tumors can invade nearby tissues. They can also break away and spread to other parts of the body.

Commonly used terms in this area are **carcinoma**, **malignancy**, **neoplasms**, **tumor**. Tumors may be benign (abnormal cells, but not cancer, as they do not spread), or malignant (cancer). Their anatomical name is "oma."

2.2 Global Cancer Facts and Figures

According to the World Health Organization (WHO) cancer facts 2017, cancer is the second leading cause of death (heart disease is the first cause of death in most of the population), and it causes nearly one in six deaths globally. Approximately, 70% of cancer deaths are recorded in low- and middle-economic countries. This is generally due to the growth and age of the population, lifestyle, including smoking and drinking alcohol, poor diet, physical inactivity, and lack of immunity, infections, etc.

The International Agency for Research on Cancer (IARC) estimated 14.1 million new cancer cases and 8.2 million cancer deaths worldwide in 2012. The WHO estimated 8.8 million cancer deaths in 2015, which included cancer of the lung, liver, colorectal, stomach, and breast on the order of 1.690, 0.788, 0.774, 0.754, and 0.571 million, respectively. However, the IARC has expected a global cancer burden of up to 21.7 million new cancer cases and 13 million cancer deaths by 2030.

2.3 Characteristics and Causes of Cancer Cells

The characteristics of cancer cells are mentioned briefly as follows:

1. Cancer cells do not stop growing and dividing.
2. Cancer cells ignore signals from other cells.
3. Cancer cells do not stick together.
4. Cancer cells do not specialize.
5. Cancer cells do not repair themselves or die.
6. The nuclei of the cancer cells are bigger in size.

More dangerous, or malignant, tumors form when two of the following occur:

1. A cancerous cell manages to move throughout the body using the blood or lymphatic systems, destroying healthy tissue in a process called invasion.
2. That cell manages to divide and grow, creating new blood vessels to feed itself in a process called angiogenesis.

Cancer is caused by many factors, such as (1) chemicals, (2) radiations, (3) viruses, (4) chromosome replication errors, and (5) DNA misrepair. Mutants (carcinogenic agents), such as chemicals and radiations, alter the DNA by modifying nucleotide sequence of the genome. This process is known as **mutation** and it is the main source of **genetic variation** (refers to diversity in gene frequencies between individuals or populations) in addition to mechanisms such as sexual reproduction and genetic drift. The DNA altered by mutation produces error during DNA replication and in turn causes error in protein synthesis, which affects the normal cellular activities, such as cell growth, cell division, and cell aging. Several cancer viruses (human papilloma viruses—cervical cancer, hepatitis B virus—liver cancer etc.) can change cells by transferring their genetic material into the human cell DNA. These infected cells will be controlled by the viral genes, and then they function abnormally.

2.4 Types of Cancer

The major types of cancer are carcinoma, sarcoma, melanoma, lymphoma, leukemia, and central nervous system cancers. Carcinoma is a common type of cancer that originates in the skin, lungs, breasts, pancreas, other cell lining (digestive tract, respiratory tract, etc.)

and tissue lining organs and glands. It can be classified as (1) basal cell carcinoma, (2) squamous cell carcinoma, (3) renal cell carcinoma, (4) invasive ductal carcinoma, and (5) adenocarcinoma. It can either spread to other parts of the body or remain in the same place where it started.

Sarcoma is a rare type of cancer that originates in connective cells and tissues of the body, such as bones, muscles, fat, tendons, cartilage, nerves, and blood vessels. Even though there are more than 50 types of sarcoma, they are grouped into soft tissue sarcoma and bone sarcoma (also called osteosarcoma).

Melanoma (also called malignant melanoma) is a common type of cancer that originates from pigment-containing cells (melanocytes) that are commonly present in the skin and rarely in the mouth, intestine, and eye.

Lymphoma is a type of cancer that originates in infection-fighting lymphocyte cells, which are present in the lymph nodes, spleen, thymus, and bone marrow of the immune system. The two main types of lymphoma are non-Hodgkin lymphoma and Hodgkin lymphoma.

Leukemia is a type of cancer that originates in the blood-forming cells inside the bone marrow. This is caused by a rapid increase in the abnormal white blood cells that push out the normal red blood cells and platelets.

2.4.1 Central Nervous System Cancers

They are cancers that begin in the tissues of the brain and spinal cord—"brain and spinal cord tumors," gliomas, meningiomas, pituitary adenomas, vestibular schwannomas, primary CNS lymphomas, and primitive neuroectodermal tumors.

Excluded in the aforementioned sorts recorded are metastatic malignancies; this is on the grounds that metastatic tumor cells more often than not emerge from a cell sort recorded earlier and the real contrast from the earlier sorts is that these cells are currently present in a tissue from which the growth cells did not initially emerge. Thus, if the expressions "metastatic malignancy" is utilized, for exactness, the tissue from which the growth cells emerged ought to be incorporated. For instance, a patient may state they have or are determined to have "metastatic growth"; however, the more exact explanation is "metastatic" (bosom, lung, colon, or other sort) malignancy with spread to the organ in which it has been found. Another illustration is the following: A specialist depicting a man whose prostate disease has spread to his bones ought to state that the man has a metastatic spread of prostate tumor to the bone. This is not "bone tumor," which would be malignancy that began in the bone cells. Metastatic prostate growth to bone is dealt with uniquely in contrast to spread of lung disease to the bone.

2.5 Cancer Stem Cell Theory

Cancer stem cell theory is gaining popularity among biologists toward cancer research. The following implications have been derived over the past few decades (Reya et al., 2001; Han et al., 2013).

1. All solid and liquid cancers consist of different types of cells.
2. Among all cancerous cells, a few (<1%) behave as cancer stem cells.

3. Cancer stem cells may arise from normal stem cells, precursor cells (cells formed from normal stem cells), or more differentiated cells through multiple mutations of genes.
4. Due to the genetic and epigenetic instability of these cells, mutations are accumulated that initiate the ability for self-renewal and carcinogenesis in the corrupted stem cells.
5. Like normal stem cells, the cancer stem cells have the capacity to reproduce themselves (self-renewal), produce differentiated progeny, express specific surface markers and oncogenes, follow similar signaling pathways, and sustain the cancer by generating heterogeneous lineages of cancer cells.
6. Unlike normal stem cells, cancer stem cells are rare cells with infinite potential for self-renewal that can form new tumors.
7. Cancer stem cells are critical cells responsible for tumor recurrence as it acts as reservoir of cancer cells, therapeutic resistance, and even metastasis.

2.6 Tumor Microenvironment

There are three major processes involved in cancer initiation and progression. They are: (1) mutations, (2) epigenetic changes (alterations in gene expression), and (3) changes in the tumor microenvironment. The tumor microenvironment can be divided into the chemical microenvironment and the cellular microenvironment (Subarsky and Hill, 2003).

The chemical microenvironment of tumors includes pH, concentration of molecules, such as O_2, PO_2, NO, etc., and metabolites like glucose, glutamine, lactate etc. Generally, solid tumors contain hypoxic (with poor oxygenation) and acidic chemical microenvironments. Many studies confirmed that conditions of hypoxia alter the gene expression. These hypoxia-induced alterations in gene expression modify nonspecific stress responses, anaerobic metabolism (metabolism without O_2), angiogenesis by promoting margination and extravasation of circulating tumor cells, tissue remodeling, and cell–cell contacts by inducing collagen hydroxylation (introduction of OH group into collagen) that stiffen the extracellular matrix. In the meanwhile, the hypoxia acts as a driving force on the cellular microenvironment to enhance cancer progression (Gregg, 2016).

The cellular microenvironment of a tumor consists of tumor cells, stromal cells, and its extracellular matrix (ECM). Tumor cells consist of cancer stem cells. Stromal cells are connective tissue cells of any organ and include fibroblasts, smooth muscle cells, adipocytes, vascular epithelial cells, lymphatic epithelial cells, bone marrow–derived cells, neurophils, mesenchymal stem cells, etc. When these stromal cells are activated by tumor cells, the process of metastasis is promoted with the involvement of the ECM.

2.7 Carcinogenesis

Induction of cancer by genetic mutations, which occur due to chemical and radiation exposure, abnormality in metabolism, etc., is referred to as carcinogenesis. The basic information about carcinogenesis is discussed in the following text (Alberts et al., 2007; Hanahan and Weinberg, 2011; Weinberg, 2013).

2.7.1 Cellular Basis of Carcinogenesis

As discussed in Chapter 1, all normal cells divide when they progress through the cell cycle in a regular manner. The synthesis of the required biomolecules for cell division is checked, corrected, verified, and regulated at various checkpoints. These checkpoints are controlled by specific proteins and enzymes, which is described as follows:

1. The master protein kinases, namely cyclin-dependent kinases (CDKs), control the progression of cells through the cell cycle by phosphorylation and also by activating other lower level enzymes. CDKs are activated by their subunits called cyclins, and their complexes are further driven by CDK inhibitors.
2. The most important checkpoint in the cell cycle is the restriction point (R point or the commitment point). It takes a decision about the activation of S-phase CDKs for DNA synthesis in order to allow the cells to progress through their cycle. However, the R point is controlled by a key regulatory transcription protein present in the nucleus, which is influenced by the extracellular mitogenic signals.
3. The guardian of the genome, namely the p53 protein, either activates the cell cycle or arrests it at all the checkpoints until the cell repairs its damages or induces apoptosis (programmed cell death) if the damage is not repairable.

Further, all the normal cells complete a definite number of cycles in their life before senescence (discussed in Chapter 3). This is achieved by the degradation of telomeres (end of the chromosome) of the cell based on the number of cycles that has been completed. These telomeres are controlled by the enzyme telomerase.

Aberrant activation of CDKs, regulatory proteins, and p53 proteins (>50%) leads to irregularities in the cell cycle. Further, activation of telomerase (~90% contribution) triggers the cell to achieve an infinite life (referred to as immortalization). This also happens when the cell repairs its dicentric chromosome using a non-homologous end-joining mechanism (discussed in Chapter 3). Hence, a normal cell is transformed into a malignant cell due to its irregular cell cycle, failure in apoptosis, and cell immortalization.

2.7.2 Cell Signaling in Carcinogenesis

In normal cells, growth factors (GFs) are involved in the process of maintaining tissue homeostasis (the property of a system in which the pH, temperature, concentration of ions, etc. are maintained at the same level), and transmit growth signal from one cell to another. To do so, these GFs (e.g., mitogen) activate the GFR (mitogen receptor), which, in turn, activates many signal-transducing proteins (e.g., Ras) present in the interior of the cell membrane. Active Ras activates MAP kinase signaling pathway that activates several target proteins (e.g., transcription factor). This again activates the Myc gene, which controls cell cycle regulatory genes.

In cancer cells, this pathway is aberrated when the GFR gene is mutated. Hence, the cell starts to activate the Ras proteins, which continuously pass the growth signal to the cell cycle regulatory genes even in the absence of GFs. This abnormal growth in the kinesis pathway simulates the cells to proliferate rapidly, disobeying their normal external signals, and finally transforms the normal cells into malignant cells.

2.7.3 Genes Involved in Carcinogenesis

Genes involved in the process of carcinogenesis are broadly classified into three. They are oncogenes (cell accelerators), tumor suppressor genes (cell decelerator), and DNA repair genes.

All the normal cells consist of proto-oncogenes (>20 in human genome) to regulate the cell proliferation and differentiation when the GF is activated. If the cell is subjected to point mutation and abnormal gene amplification, the growth factors, replication signals, and transcription factors are rapidly synthesized, which leads to abnormality in the MAP kinase signaling pathway. It converts the proto-oncogene into an oncogene, which can promote aberrant cell divisions, independent of growth signals, for example, the RAS oncogene, which is responsible for most of the human cancers.

The tumor suppressor gene (>16 in human genome) controls several proteins that are responsible for (1) suppressing cell proliferation, (2) reverse regulators of growth signaling pathways and cell cycle progress, (3) regulating cell cycle checkpoint proteins, (4) regulating proteins that induce apoptosis, and (5) regulating proteins involved in the DNA repair mechanism, for example, the retinoblastoma (RB) gene, p53 gene, and APC (adenomatous polyposis coli) gene. An abnormal function or loss of function of one or both of the tumor suppressor genes (p53 into TP53-tumor protein 53) triggers abnormalities in the cell.

As mentioned in Chapter 3, DNA damage occurs frequently in all the cells either by the interaction of free radicals present inside the cell or by environmental agents. However, the cell has several effective repair mechanisms to detect, correct, and verify the damage in order to maintain the genome. Studies have reported that the human genome consists of more than 130 DNA repair genes (Wood et al., 2001) to correct the DNA damage through different repair mechanisms. When the repair gene is mutated, the normal cell starts to become a malignant cell in multiple steps.

2.7.4 Multistep Nature of Carcinogenesis

Many epidemiological studies have confirmed that the induction of cancer is a complex, multistep, and micro-evolutionary process due to the sequential acquisition of mutations in genes, which controls normal cellular activities. It is clearly explained in colorectal cancer as a three-step process. The three major steps in carcinogenesis are: (1) malignant transformation, (2) invasion, and (3) metastasis. Each step consists of a number of alterations in the cellular genome.

Malignant transformation is a continuous process occurring due to multiple genetic changes. It is estimated that a minimum of 4–7 mutated genes are required for neoplastic transformation. In this transformation, the cell generates mitogentic signals autonomously, is irresponsive to antigrowth signals, gets resistivity to apoptosis, stops differentiation, and also gets immortality power through various steps as follows. In the case of colorectal cancer, it involves many tumor suppressor genes, namely, APC, 18q TSG, p53, etc., proto-oncogenes (e.g., K-ras), and loss of functional group of DNA (e.g., DNA hypomethylation).

1. Transformation of normal epithelial tissues into hyperproliferative epithelium due to the loss of one of its APC tumor suppressor genes.
2. Transformation of hyperproliferative epithelium into early adenoma due to the addition of one more epigenetic abnormalities, DNA hypomethylation, etc.
3. Transformation of early adenoma into intermediate adenoma due to the mutation of the K-ras proto-oncogene.

4. Transformation of intermediate adenoma into dysplastic adenoma when the cell loses another tumor suppressor gene, 18q TSG.
5. Transformation of dysplastic adenoma into epithelial cancer (carcinoma) when the cell loses its important tumor suppressor gene, p53.

Invasion (displacement of carcinoma cells) consists of the following four steps:

1. Detachment of the carcinoma cells from other tumor cells by reducing the cell–cell adhesion molecules.
2. Sticking to the extracellular matrix of surrounding connective tissues and the membrane of blood vessels.
3. Degrading the extracellular matrix molecules such as collagen, glycoproteins, and proteiglycans by producing integrin molecules.
4. Movement of the cells toward the growth factors in rich places like blood/lymph vessels. This is referred to as intravasion.

Metastasis is defined as the spreading of carcinoma cells to remote places. It reduces the curability rate rapidly. It can spread through (1) natural cavities viz. peritoneum, pleura, etc., (2) lymphatic vessels, and (3) veins (*not through arteries*). Metastasis takes the following steps after their invasion:

1. Survival of the carcinoma cells in circulation.
2. Arrests the capillary bed at a distant organ site.
3. Penetration through the lymphatic or blood vessel walls followed by the growth of disseminated tumor cells. This is referred to as extravasation.
4. Formation of micrometastasis.
5. Colonization of metastatic cells.
6. Angiogenesis—Creation of new blood vessels due to the release of many different proteins and molecules from metastatic cells.
7. Growth of metastatic cells to a detectable size.

2.7.5 Metastasis Pathways

It is well known that the metastasis (cancer cell spread or dissemination) is the cause of most of the cancer deaths. As it is a complex and multistep biological process controlled by many genes, the exact mechanism of metastasis remains unclear in spite of many efforts. Dissemination of cancers may occur through one of three pathways: (1) direct seeding of body cavities or surfaces, (2) lymphatic spread, and (3) hematogenous spread.

Seeding of body depressions and surfaces may happen at any point where a threatening neoplasm infiltrates a characteristic "open field." Most regularly included is the peritoneal cavity; however, some other cavity pleural, pericardial, subarachnoid, and joint space might be influenced. Such seeding is especially normal for carcinomas emerging in the ovaries, when, not occasionally, all peritoneal surfaces get distinctly covered with an overwhelming layer of dangerous coating. Strikingly, the tumor cells may remain limited to the surface of the covered stomach viscera without entering into the substance. In some cases, bodily fluid emitting appendiceal carcinomas fills the peritoneal hole with a coagulated neoplastic mass alluded to as pseudomyxoma peritonei.

Transport through lymphatics is the most well-known pathway for the underlying dispersal of carcinomas, and sarcomas may likewise utilize this course. Tumors do not contain useful lymphatics, yet lymphatic vessels situated at the tumor edges are evidently adequate for the lymphatic spread of tumor cells. The accentuation on lymphatic spread for carcinomas and hematogenous spread for sarcomas is misdirecting, in light of the fact that at last there are various interconnections between the vascular and the lymphatic frameworks. The example of lymph hub inclusion takes after the characteristic courses of lymphatic seepage. Since carcinomas of the bosom more often than not emerge in the upper external quadrants, they for the most part spread first to the axillary lymph hubs.

Diseases of the inward quadrants deplete to the hubs along the inner mammary supply routes. From thereon the infraclavicular and supraclavicular hubs may get distinctly included. Carcinomas of the lung emerging in the major respiratory sections metastasize first to the perihilar tracheobronchial and mediastinal hubs. Nearby lymph hubs, nonetheless, might be circumvent supposed "skip metastasis" in view of venous-lymphatic anastomoses or in light of the fact that aggravation or radiation has pulverized lymphatic channels. In bosom growth, deciding the inclusion of axillary lymph hubs is critical for surveying the future course of the ailment and for choosing reasonable helpful techniques.

To maintain a strategic distance from the impressive surgical horribleness related with full axillary lymph hub dismemberment, a biopsy of sentinel hubs is frequently used to survey the nearness or nonappearance of metastatic sores in the lymph hubs. A sentinel lymph hub is characterized as "the principal hub in a territorial lymphatic bowl that gets lymph spill out of the essential tumor." Sentinel hub mapping should be possible by infusion of radiolabeled tracers and blue colors, and the utilization of solidified area upon the sentinel lymph hub at the season of surgery can direct the specialist to the proper treatment. Sentinel hub biopsy has likewise been utilized for distinguishing the spread of melanomas, colon malignancies, and different tumors (Carlos S.—Metastasis). In many cases, the territorial hubs fill in as successful hindrances to further spread of the tumor, in any event for some time. It is possible that the cells, after capture inside the hub, might be wrecked by a tumor-specific resistant reaction. Seepage of tumor-cell flotsam and jetsam or tumor antigens, or both, additionally initiates receptive changes inside hubs. Therefore, broadening of hubs might be brought about by (1) the spread and development of disease cells, or (2) responsive hyperplasia. In this manner, nodal growth in close proximity to a tumor, while it must stimulate doubt, does not really mean dispersal of the essential sore.

Hematogenous spread is run-of-the-mill in sarcomas, but in contrast, is seen in carcinomas. Courses, with their thicker dividers, are less promptly infiltrated than veins. Blood vessel spread may happen, be that as it may, when tumor cells go through the aspiratory hairlike beds, or pneumonic arteriovenous shunts, or when aspiratory metastases themselves offer ascent to extra tumor emboli. In such vascular spread, a few elements impact the examples of dissemination of the metastases. With venous intrusion the blood-borne cells follow the venous stream depleting the site of the neoplasm, and the tumor cells regularly stop in the main fine-bed they experience. Naturally, the liver and lungs are most every now and again required in such hematogenous scattering, since all entry region waste streams to the liver and all caval blood streams to the lungs.

Diseases emerging in closeness to the vertebral segment regularly embolize through the paravertebral plexus, and this pathway is included in the incessant vertebral metastases of carcinomas of the thyroid and the prostate. Certain tumors have an affinity for the attack of veins. Renal cell carcinoma frequently attacks the branches of the renal vein and after that the renal vein itself to develop in a snakelike manner up the second rate vena cava, here and there achieving the correct side of the heart. Hepatocellular carcinomas regularly enter the gateway and hepatic radicals to develop inside them into the principal venous channels. Strikingly, such intravenous development may not be joined by across-the-board spread. Histologic proof of the entry of small vessels at the site of the essential neoplasm is clearly an unfavorable component. Such changes, be that as it may, must be seen guardedly in light of the fact that, for reasons talked about later, they do not show the inescapable advancement of metastases.

Numerous perceptions propose that unimportant anatomic restriction of the neoplasm and normal pathways of venous waste do not completely clarify the systemic dissemination of metastases. For instance, bosom carcinoma specially spreads to bone, bronchogenic carcinomas have a tendency to include the adrenals and the cerebrum, and neuroblastomas spread to the liver and bones. On the other hand, skeletal muscles and the spleen, notwithstanding the expansive rate of blood stream they get and the huge vascular beds present are the site of auxiliary stores once in a while.

Based on the knowledge on genomic changes in tumor cells, it is recognized that the changes in the cell phenotype between the epithelial states (epithelial cells are well polarized and closely bound to each other by tight junctions, gap junctions, and adherent junctions) and mesenchymal states (mesenchymal cells are less polarized, loosely bound, and communicate with each other through focal points) are also considered to be some of the basic processes underlying cancer cell dissemination. These changes are expressed as epithelial mesenchymal transition (EMT) and mesenchymal epithelial transition (MET).

Generally, EMT has been categorized into three types. They are types I, II, and III transitions. Type I transition is responsible for the development of many tissues and organs during embryogenesis. Type II transition is responsible for fibrosis and wound healing. Type III transition induces cancer due to change in the gene expression of the stem cells. The normal gene expression can be changed due to growth factor–mediated crosslinking between signaling pathways, transcription factor–mediated depression, down-regulation or silencing of E-cadherin, change in the expression of microRNAs, etc.

During type III transition, cancer cells can activate local stromal cells, and recruit endothelial and mesenchymal progenitors and inflammatory cells. It triggers the proliferation of cancer cell and invasion by the secretion of additional growth factors and proteases and promotion of EMT. EMT occurs all along the tumor–host interface of carcinomas, supporting the notion that the environment triggers EMT at the tumor–host interface.

Subsequently, the disseminated mesenchymal tumor cells undergo the reverse transition, MET, at the site of metastases, as metastases recapitulate the pathology of their corresponding primary tumors. This may due to the up-regulation or reexpression of E-cadherin and the acquisition of differentiated epithelial cell features. EMT is thought to be critical for the initial transformation from benign to invasive carcinoma, whereas MET (the reverse of EMT) is critical for the later stages of metastasis.

2.8 Cancer as a Genetic Disease

Cancer arises due to the mutation of genes in the cells. Mutation can occur in genes that are present in both somatic and reproductive cells. If the genes in the somatic cells of a person get mutated, it appears to be a somatic disease during his lifetime. Nevertheless, the mutation that occurs in reproductive cells when it passes through various phases of the cell cycle and during meiosis will pass on to his progeny/offspring/generation and becomes a genetic disease. Hence, cancer is also considered to be a genetic disease.

2.9 Classification of Cancer

1. Based on the primary site of origin

   Cancers may be of specific types like breast cancer, lung cancer, prostate cancer, liver cancer, renal cell carcinoma (kidney cancer), oral cancer, brain cancer, etc.

2. Based on the tissue type

   Carcinoma—Originates from the epithelial layer of cells, for example, squamous cell carcinoma, adenocarccinoma, transitional cell carcinoma, and basal cell carcinoma.

   Sarcoma—Originates in connective and supportive tissues, for example, bone, soft tissue, cartilage, and muscle sarcoma

   Myeloma—Originate in the plasma cells of bone marrow

   Leukemia—A group of blood cancers

   Lymphoma—Cancers of the lymphatic system

   Mixed types—Cancers that have two or more specific components of cancer

3. According to the grade

   The abnormality of the cells with respect to surrounding normal tissues determines the grade of the cancer. Increasing abnormality increases the grade from 1 to 4.

   Grade 1—Cells are well differentiated with slight abnormality

   Grade 2—Cells are moderately differentiated and slightly more abnormal

   Grade 3—Cells are poorly differentiated and very abnormal

   Grade 4—Cells are immature, primitive, and undifferentiated

4. According to their stage

   The most commonly used method uses the classification in terms of tumor size (T), the degree of regional spread or node involvement (N), and distant metastasis (M). This is called the **TNM** staging as given here:

   T—Primary tumor

   TX—Primary tumor cannot be assessed

   T0—No evidence of primary tumor

   T1—Solitary tumor 2 cm or less in the greatest dimension without vascular invasion

   T2—Solitary tumor 2 cm or less in the greatest dimension with vascular invasion; or multiple tumors limited to one lobe, none more than 2 cm in the greatest dimension without vascular invasion; or solitary tumor more than 2 cm in the greatest dimension without vascular invasion.

   T3—Solitary tumor more than 2 cm in the greatest dimension with vascular invasion; or multiple tumors limited to one lobe, none more than 2 cm in the greatest dimension with vascular invasion; or multiple tumors limited to one lobe, any more than 2 cm in the greatest dimension with or without vascular invasion.

T4—Multiple tumors in more than one lobe; or tumor(s) involve(s) a major branch of the portal or hepatic vein(s); or tumor(s) with the direct invasion of adjacent organs other than gallbladder; or tumor(s) with perforation of visceral peritoneum.

N—Regional lymph nodes

NX—Regional lymph nodes cannot be assessed

N0—No regional lymph node metastasis

N1—Regional lymph node metastasis

M—Distant metastasis

MX—Distant metastasis cannot be assessed

M0—No distant metastasis

M1—Distant metastasis

Stage grouping:

- Stage I—T1 N0 M0
- Stage II—T2 N0 M0
- Stage IIIA—T3 N0 M0
- Stage IIIB—T1 N1 M0 T2 N1 M0 T3 N1 M0
- Stage IVA—T4 Any N M0
- Stage IVB—Any T Any N M1

2.10 Methods to Diagnose Cancer

The various methods and techniques to diagnose (detect) cancer are listed as follows:

1. The history of symptoms is the first step in diagnosing cancer.
2. Biopsy—Based on the examination of a tissue sample taken from a cancerous region.
   a. *With a needle*: The doctor uses a needle to withdraw tissue or fluid.
   b. *With an endoscope*: The doctor looks at areas inside the body using a thin, lighted tube called an endoscope. The scope is inserted through a natural opening, such as the mouth. Then, the doctor uses a special tool to remove tissue or cells through the tube.
   c. *With surgery*: Surgery may be excisional or incisional.
      i. In an excisional biopsy, the surgeon removes the entire tumor. Often some of the normal tissue around the tumor also is removed.
      ii. In an incisional biopsy, the surgeon removes just part of the tumor.
3. Various imaging modalities, such as x-rays, CT, MRI, ultrasound, PET, nuclear scan etc.
4. *Urine culture*: a test to identify germs that may be causing a urinary tract infection (UTI).
5. Endoscopy, etc.

2.11 Cancer Treatment

2.11.1 Various Treatment Modalities

Cancer treatment uses different modalities either alone or in combination. The increasing use of multimodality treatment has blurred the once-clear rules of each treatment modality. The combination of different treatment modalities now offers a more effective cure and reduction in normal-tissue toxicity. Traditionally, cancer treatment modalities include

- Surgery (for local and local–regional disease)
- Radiation therapy (for local and local–regional disease)
- Chemotherapy (for systemic disease)

Other important methods include the following:

- Hormonal therapy (for selected cancers, e.g., prostate, breast, endometrium)
- Immunotherapy (monoclonal antibodies, interferons, and other biologic response modifiers and tumor vaccines)
- Targeted drugs that exploit the growing knowledge of cellular and molecular biology

Overall treatment should be coordinated among a radiation oncologist, a surgeon, and a medical oncologist, where appropriate. Choice of modalities is constantly evolving, and numerous controlled research trials continue. When available and appropriate, clinical trial participation should be considered and discussed with patients.

Various terms are used to describe the response to treatment. The disease-free interval often serves as an indicator of cure and varies with cancer type. For example, lung, colon, bladder, large cell lymphomas, and testicular cancers are usually cured if a 5-year disease-free interval occurs. However, breast and prostate cancers may recur long after 5 years, an event defining tumor dormancy (now a major area of research); thus, a 10-year disease-free interval is more indicative of cure.

Often, modalities are combined to create a treatment program that is appropriate for the patient and is based on patient and tumor characteristics as well as patient preferences.

Individualized cancer treatment, which is the goal of the next revolution in radiation oncology, can be achieved with intense approaches in the fields of DNA repair, cell cycle control and signaling transduction, tumor microenvironment, and molecular targeting in radiation oncology.

2.11.2 Milestones in Cancer Treatment

Radiation therapy can cure many cancers (see 5-Year Disease-Free Survival Rates by Cancer Therapy), particularly those that are localized or that can be completely encompassed within the radiation field. Radiation therapy plus surgery (for head and neck, laryngeal, or uterine cancers) or combined with chemotherapy and surgery (for sarcomas or breast, esophageal, lung, or rectal cancers) improves cure rates and allows for more limited surgery as compared with traditional surgical resection. Radiation cannot destroy malignant cells without destroying some normal cells as well. Therefore, the risk to normal tissue must be weighed against the potential gain in treating the malignant cells. The final outcome of a dose of radiation depends on numerous factors, including the nature of the delivered radiation (mode, timing, volume, and dose) and the properties of the tumor (cell cycle phase, oxygenation, molecular properties, and inherent sensitivity to radiation). Treatment is tailored to take advantage of the cellular kinetics of tumor growth, with the aim of maximizing damage to the tumor while minimizing damage to normal tissues.

Radiation therapy can provide significant palliation when a cure is not possible (palliative radiotherapy):

- *For brain tumors*: Prolongs patient functioning and prevents neurologic complications
- *For cancers that compress the spinal cord*: Prevents progression of neurologic deficits
- *For superior vena cava syndromes*: Relieves venous obstruction
- *For painful bone lesions*: Usually relieves symptoms
- *For advanced lung cancer*: Usually relieves symptoms, such as hemoptysis, cough, dyspnea, and medial obstruction

Improvements in radiotherapy over the last two decades include technical advances in instrumentation, software treatment planning, and diagnostic techniques, enabling a more accurate dose delivery to the tumor while minimizing the dose to surrounding healthy tissue.

2.11.3 Significance of Tumor Characteristics in Treatment and Prognosis

Tumor factors that govern both treatment and prognosis include those relate to pathology, including histologic type, grade, growth pattern, or pattern of invasion. One of the most important prognostic factors in cancer is the anatomic disease extent that routinely is classified according to the TNM classification. Additional factors describing the anatomic extent of disease include bulk of tumor, site of tumor, number of separate lesions, and levels of tumor markers such as a-fetoprotein, CA125, prostate-specific antigen, b-human chorionic gonadotropin, etc. A different group of tumor-related prognostic factors are those that characterize tumor biology and include proliferation indices, percent in S-phase, MIB-1, Ki-67, etc. (Jansen et al., 1998; Holte et al., 1999). Hormone receptors and molecular and genetic markers are also elements of the tumor-related prognostic factors. Finally, the presence of symptoms, local or systemic, is a powerful tumor-related factor; however, it can often be mistaken as a patient-related factor.

2.12 Radiation in Cancer Treatment

2.12.1 Radiobiology of Tumor Response to Radiotherapy

Radiobiological parameters that determine tumor response to radiotherapy include intrinsic radiosensitivity, proliferation rate, and hypoxia. These three factors are considered separately and the goal is to measure them all to maximize the chance of accurately predicting response. Each of these factors is controlled by many genes and pathways. Thus, to maximize the chance of reliably predicting the success of a treatment in an individual, multiple factors, or multiple genes, need to be measured, both in the tumor and in the patient.

2.12.1.1 Tumor Proliferation

The importance of tumor proliferation is most clearly shown by the higher doses required to control a tumor when overall treatment is increased (first demonstrated by Withers et al., 1988). Further evidence comes from studies showing loss of local tumor control as a result of gaps in treatment whether planned or unplanned. There is also increasing evidence from randomized trials that accelerated regimes can improve outcome (Horriot et al., 1997; Saunders et al., 1999; Bourhis et al., 2006).

Methods for measuring tumor proliferation include measurement of mitotic index, the proportion of cells in the S phase of the cell cycle, labeling index, tumor potential doubling time, and proliferation antibodies that measure growth fraction.

2.12.1.2 Tumor Hypoxia

Over the past 50 years, considerable evidence has shown the importance of hypoxia in tumors as a factor limiting the success of radiotherapy. There is now also good evidence that hypoxia plays a key role in tumor progression by promoting both angiogenesis and metastasis. Pioneers in hypoxia predictive assay research (Vaupel et al., 1991; Hockel et al., 1993) and many other independent groups have shown that pretreatment measurements of tumor oxygenation were a significant prognostic factor in a variety of cancers. A very important finding in this area of research was the demonstration that hypoxic tumors respond poorly not only to radiotherapy but also when treated with surgery alone. There is therefore considerable interest in assessing a methodology for measuring tumor hypoxia for widespread clinical use. Such methods include the Eppendorf pO_2 histograph, hypoxia-specific chemical probes such as pimonidazole (Raleigh et al., 2000; Kaanders et al., 2002; Nordsmark et al., 2005), and EF5 (Evans et al., 2006). A variety of methods have been used to score vascularity (West et al., 2001), because of the known association between hypoxia and angiogenesis. Finally, non-invasive imaging is another approach being investigated, including positron emission tomography (PET) (Krause et al., 2006), computed tomography (CT), magnetic resonance spectroscopy imaging (MRS), and magnetic resonance imaging (MRI).

2.12.1.3 Tumor Radiosensitivity

There is strong evidence, from cell lines, animal tumor models, and in the clinic, for the importance of the wide variation in intrinsic radiosensitivity even between tumors of similar origin and histological type. First attempts assessing radiosensitivity of human tumors mainly refer to clonogenic assays, usually specifying the surviving fraction at 2 Gy (SF2), but unfortunately showed a poor success rate for human tumors and need time of 1–4 weeks (West et al., 1997; Bjork-Eriksson et al., 2000). Alternative assays include measuring chromosome and DNA damage (Coco Martin et al., 1999; Klokov et al., 2006). With the advent of new technologies for the simultaneous analysis of the expression of the whole human genome, there is also interest in the application of microarray methods for obtaining profiles that predict tumor-cell sensitivity (Torres-Roca et al., 2005). In fact, even though no reliable method for human tumor radiosensitivity measurement has been yet established, all the assays mentioned here have confirmed the key role of radiosensitivity underlying differences in response to radiotherapy.

The future of radiation oncology lies in exploiting the genetics or the microenvironment of the tumor. The IAEA reported (West et al., 2005) the ongoing studies worldwide aiming to characterize the molecular profiles that predict normal-tissue and tumor-tissue radio response. Among those, the GENEPI project (genetic pathways for the prediction of the effects of radiation) launched by the European Society for Therapeutic Radiology and Oncology (ESTRO) is the most comprehensive one.

2.12.1.3.1 Combined Radiotherapy and Chemotherapy

Combined chemotherapy and radiotherapy might refer to sequential association or to concomitant association. Although several mechanisms of interaction between drugs and radiation have been identified (modulation of DNA and chromosome damage and repair, cell cycle synchronization, enhanced induction of apoptosis, and re-oxygenation), in a clinical setting it is most likely that a key benefit is the inhibition of tumor-cell proliferation by drugs during the radiation interfraction interval.

Concomitant administration of chemotherapy and radiation gives increased early normal-tissue toxicity due to inhibition of stem-cell or precursor-cell proliferation. Late normal-tissue damage is likely to be enhanced through inhibition of DNA repair, and by specific mechanisms of drug toxicity in sensitive tissues. Several randomized trials with concomitant chemoradiotherapy have been conducted in most cancer sites showing a significant increase in loco-regional control in many disease sites with a consequent improvement in patient survival.

A drug may sensitize the radiation or may kill cells by independent means. Alternatively, a drug may inhibit cellular repopulation or act as a cytoprotector. A limited number of publications (e.g., Jones et al., 2005; Plataniotis and Dale, 2007) have presented drug mechanisms mathematically in order to estimate the equivalent radiation effect of a drug. In fact if cytotoxic drug effects could be expressed in terms of equivalent biologically effective dose of radiation then relative contributions of radiation and chemotherapy in combined treatments could be assessed and consequently optimum schedules could be designed (Dale and Jones, 2007).

An ideal global model of tumor control in an attempt to simulate clinical reality would incorporate the effects of radiation dose, fractionation, hypoxia, blood flow, and concomitant drug therapy.

2.12.2 Effects of Signaling Abnormalities on Radiation Responses

The predominant mechanism by which ionizing radiation induces cell death in mammalian cells is thought to be the reproductive model of cell death: DNA single- and double-strand breaks are produced by incident photons, electrons, or free radicals (radiation induced). However, the cellular radiation response to radiation is more complex than simply DNA damage repair. It is now evident that DNA damage by irradiation induces a complex network of inter- and intracellular signaling that can lead either to cell cycle arrest and induction of the DNA repair machinery or to programmed cell death in distinct cells. But ionizing radiation not only induces signaling pathway modifications upon DNA damage, it also induces specific signaling pathway modifications at the cell membrane (e.g., growth factor activation) or in the cytoplasm independent of DNA damage, as a stress response upon membrane alterations and membrane receptor activation. Receptor tyrosine kinases (RTKs) make up a group of receptors that are critically involved in human cancer. Examples of receptor tyrosine kinases are the insulin receptor, the epithelial growth factor receptor (EGFR) family, and the vascular endothelial growth factor receptor (VEGFR) family.

Mutations or pharmacological modulation of key elements of this complex regulatory network of signal transduction cascades can dramatically change the radiation response of tumor cells, for example, from DNA repair to cell death and thus change their biologic radiosensitivity.

Different cancers behave and respond to treatment differently. Therefore, we need to understand the molecules involved and their functions, such as intracellular signaling cascades that regulate radiation sensitivity and resistance of tumor cells and normal cells. As basic molecular biology techniques have been progressing rapidly and the human genome project has completed mapping and sequencing all the genes in humans, the information gained will greatly impact cancer treatment.

Objective-Type Questions

1. Cell gets infinite life by the activation of
   a. CDKs
   b. Regulatory proteins
   c. p53 proteins
   d. Telomeres
2. The following is the reverse regulator of growth signaling pathways and cell cycle progress.
   a. Proto-oncogene
   b. tumor suppressor gene
   c. DNA repair gene
   d. All of the above
3. Cancer cells that are immature, primitive, and undifferentiated are classified into the following grade.
   a. Grade 1
   b. Grade 2
   c. Grade 3
   d. Grade 4
4. Papilloma virus can induce the following cancer.
   a. Cervical cancer
   b. Liver cancer
   c. Lung cancer
   d. None
5. Factor/s that describe the anatomic extent of the disease is/are
   a. Bulk of tumor
   b. Number of separate lesions
   c. Levels of tumor markers
   d. All

6. Peak incidence of second cancers in radiotherapy patients occurs
   a. 1–5 years
   b. 5–10 years
   c. 5–20 years
   d. >20 years
7. Final outcome of a dose of radiation depends on
   a. Mode of treatment
   b. Cell cycle phase
   c. Inherent radio sensitivity
   d. All
8. The dimension of the T3 solitary tumor is
   a. <1 cm
   b. 1–2 cm
   c. >2 cm
   d. Multiple tumors
9. p53 is a
   a. Proto-oncogene
   b. tumor suppressor gene
   c. DNA repair gene
   d. None
10. Bone cancer is a
    a. Carcinoma
    b. Myeloma
    c. Sarcoma
    d. Lymphoma

3

Interaction of Radiation with Cells

Objective

When a cell is exposed to radiation, it undergoes various interactions and modifications at different levels before its death or abnormality. The processes and mechanisms involved during these interactions are discussed in this chapter.

3.1 Concepts of Microdosimetry

It is well known that the emission of radiation is stochastic (random) in nature. Hence, its energy transfer and deposition also become a stochastic process. However, the conventional treatment/dosimetry is applied to treat macroscopic regions using high flux radiation based on the absorbed dose (the mean energy imparted to unit mass of matter by ionizing radiation), particle-stopping power (average energy loss per unit path length), and linear energy transfer (LET, average energy transferred per unit path length). The necessary conditions to apply these quantities are (1) homogeneous distribution of energy depositions in uniformly irradiated matter and (2) secondary particle equilibrium. Under these conditions, the radiation effects are proportional to the absorbed dose and the standard deviation is low in all the processes. But, these conditions may not be satisfied when we deal with small doses (mGy), with densely ionizing radiation, and its energy deposition at the cellular level (most of the animal cells are in the range of 10–30 μm). It necessitates the need for the use of microdosimetry to connect the amount of energy deposited in cells and its biological effects. It is also an effective way of measurement to determine the beam quality.

Microdosimetry is referred to as the study of spatial (360° variation), temporal (spatial distribution with time), and spectral aspects (energy and type of radiation) of the stochastic nature of the energy deposition processes in microscopic structures. However, the amount of detail needed to predict tissue response in radiotherapy remains unclear.

As discussed in Report No. 1914/D based on ICRU 36, 1984, the basic quantities used in microdosimetry are briefly given here.

1. *Energy deposit*: The energy deposited by an ionizing particle in a single interaction at the transfer point (the point where interaction takes place) is called energy deposit, ε_i.

$$\varepsilon_i = E_{in} - E_{out} - Q_{\Delta m}$$

   where
   E_{in} is the incident energy of the particle
   E_{out} is the sum of the energies of all ionizing particles leaving the interaction
   $Q_{\Delta m}$ is the energy involved in the change of the rest mass of the atom and all particles involved in the interaction

2. *Energy imparted*: The track of the particle that can be superimposed on a volume of interest V, is called a sensitive volume, target, or a sensitive site. The contribution from all energy deposits ε_i in the volume V is called the energy imparted, ε (J).

$$\varepsilon = \sum_i \varepsilon_i$$

3. *Specific energy*: The specific energy (z) replaces the absorbed dose in microdosimetry. It is defined as the energy imparted locally in a small volume of mass. The energy imparted to the target depends not only on the topology of the track but also on the size, shape, and composition of the target.

$$z = \frac{\varepsilon}{m}$$

4. *Lineal energy (y)*: The lineal energy (stochastic quantity) replaces the LET (statistical quantity) in microdosimetry. It is defined as the energy imparted in one event divided by the mean chord length l (target volume) that results from the random interception of the site by a straight line. The mean chord length is equal to 4V/S for a convex site of volume V and surface S.

$$y = \frac{\varepsilon}{l}$$

Lineal energy can be measured with A150 proportional counters containing tissue-equivalent gas at low pressure to simulate micrometer-sized target volumes. But, the "true" target volumes of life science are of nanometer size DNA (2.3 nm dia.). Because the greater part of radiation damage to cells starts with the initial damage to segments of the DNA, the frequency distribution of the number of particle interactions in nanometer-sized target volumes must be measured. Based on this, the microdosimetry concept is extended to nanodosimetry to measure the cluster-size distribution. The cluster size is the number of ionizations produced by a particle in a specified target volume (Figure 3.1). Even though there are many challenges, the vision of nanodosimetry is to either replace the conventional dosimetric quantity, absorbed dose by nanodosimetric quantity, or an additional quantity to quantify radiation at the DNA level.

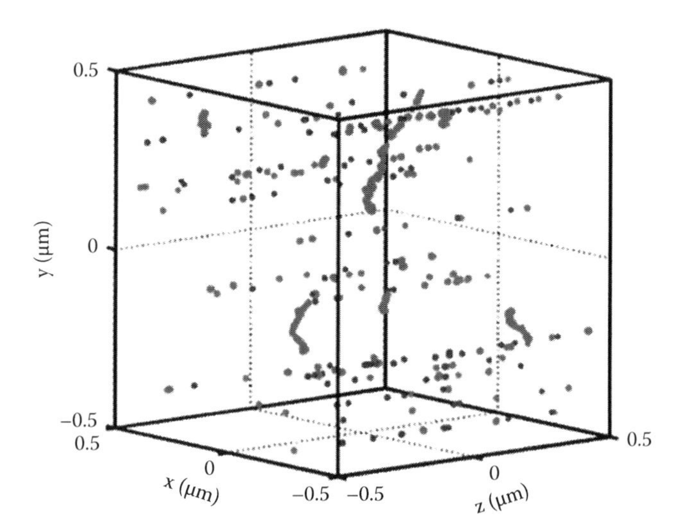

FIGURE 3.1
Geant4-DNA Monte Carlo code–simulated cluster-size distribution pattern of gamma rays.

In order to study the energy deposition processes in microscopic and nanometric structures, the interaction of radiation with cells at various levels, namely, atomic, molecular, and cellular levels, is discussed in this chapter.

3.2 Various Stages of Interaction of Radiation with Cells

In general, ionizing radiation interacts with cells in four stages as shown in Figure 3.2. They are (1) physical (2) pre-chemical or physicochemical (3) chemical, and (4) biological stages. In the physical stage, the kinetic energy of the radiation is transferred to atoms or a molecule, which leads to excitation and ionization of these molecules. It takes just 10^{-16}–10^{-15} seconds. During the pre-chemical stage, free radicals are formed in 10^{-15}–10^{-12} seconds due to absorption of the released kinetic energy. In the chemical stage, chemical reactions between free radicals and molecules take place, which leads to the formation of abnormal biomolecules (e.g., DNA). This is done in 10^{-12}–10^{-6} seconds. Due to these abnormalities, injuries are formed at all levels ranging from cells to organisms. It takes a few seconds to a few years depending on the type of radiation, energy of radiation, part of body irradiated, dose, and dose rate. Hence, the interaction of radiation with cell is characterized as a combination of physics, chemistry, and biology and it is discussed and explained here.

3.3 Interaction of Radiation with Cells at the Atomic Level

The major physics processes involved at the atomic level of radiation interaction are (1) excitation and (2) ionization at its physical stage. Excitation is the process of exciting an atomic electron from its original ground state to the excited state due to energy transfer

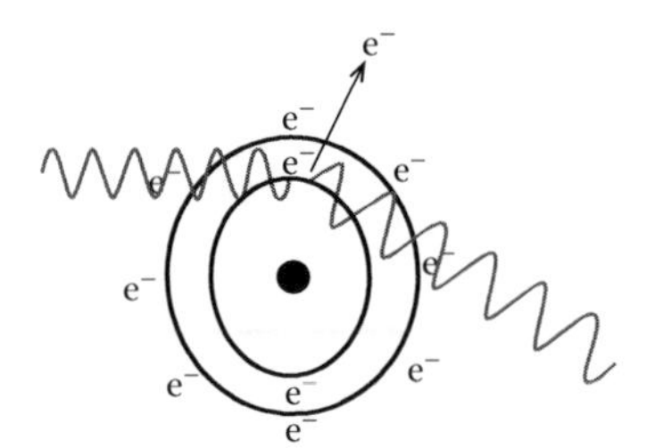

Physical stage (≤femtosecond) – Production of ions (H_2O^*, $H_2O^+ + e^-$)

Prechemical stage (femtosecond to picosecond) – Production of free radicals
($H^\bullet + HO^\bullet$, $H_2 + 2HO$, $HO^\bullet + H_3O^+$, $HO^\bullet + H_2 + OH^-$ e^-_{aq})

Chemical stage (picoseconds to microsecond) – Breakage of chemical bonds
(e^-_{aq}, $H^\bullet$, $HO^\bullet$, $HO_2^\bullet$, OH^-, H_3O^+, H_2, H_2O_2)

Biological stage (millisecond to many years)

FIGURE 3.2
Various stages of interaction of radiation with cells.

from the ionizing radiation to atomic electrons. If the transferred energy is not sufficient enough to eject this electron, the excited electron will come down to its original ground state in 10^{-8} seconds. This is known as excitation. If the transferred energy is sufficient to eject that electron, ionization takes place. Generally, ionization is referred to as the process of producing ions either by removing or adding ions (electrons). Commonly, ionizing radiation removes electrons by physical processes. For example, photons produce ion pairs (a negatively charged electron and a positively charged remainder atom) through three major physical processes, such as photoelectric effect, Compton effect, and pair production, depending upon its energy.

Absorption of energy sufficient to remove an electron can result in bond breaks.

$$\text{Ionizing radiation} + RH \rightarrow R^- + H^+$$

Excitation of atoms in key molecules can also result in bond breaks. In this case, energy can be transferred along the molecule to a side of bond weakness and cause a break. Also tautomeric shifts can occur where the energy of excitation can cause predominance of one molecule form.

$$\begin{array}{ccc} OH & & O \\ I & & II \\ R-C=NH & \leftrightarrow & R-C=NH_2 \\ \text{imidol(enol)} & & \text{amide(keto)} \end{array}$$

Imidol and amide are tautomers in equilibrium with the amide (keto) being predominant. In the amide form, the molecule reacts in the proper biochemical chain. The introduction of excitation energy shifts the equilibrium to the imidol (enol) form. This either causes no reaction or causes a misread at the active site of an enzyme in the chain.

3.4 Interaction of Radiation with Cells at the Molecular Level

All the interactions of radiation with cells are initiated due to the absorption of energy from ionizing radiation either by biomolecules (DNA, RNA, protein, enzymes, etc.) or in their suspension medium (such as water) through ionization and excitation. Depending on the location of the interaction, all radiation interactions with cells are classified as either direct action or indirect action.

3.4.1 Direct Action of Radiation

Direct action takes place when an original/initial ionizing incident happens in critical biomolecules like DNA as shown in Figure 3.3. In the process of ionization, high LET radiations can also break the covalent bond in the biomolecules and produce two free radicals that are highly reactive. It leads to inactivation or functional alteration of the biomolecule.

$$\text{High LET radiation} + RH \rightarrow RH^{+} + e^{-}$$

$$RH^{+} \rightarrow R^{\bullet} + H^{+}$$

$$R - R' \rightarrow R^{\bullet} + R'^{\bullet} \text{ (production of molecular free radicals)}$$

A free radical is an electrically neutral atom with an unpaired electron in its outermost orbit. For example, in $OH^{\bullet}$ (hydroxyl free radical), the subatomic level of oxygen (O) molecule is $1s^2$ (paired ↑↓) $2s^2$ (paired ↑↓) $2p^4$ (Px is paired ↑↓ but Py ↑ and Pz ↑ orbits are unpaired) and the hydrogen (H) molecule is $1s^1$ (↑ unpaired). So the electron in the $1s^1$ orbit of H pairs with the unpaired electron in the Py orbit of O to become $OH^{\bullet}$. The $OH^{\bullet}$ radical consists of one more unpaired electron in the Pz orbit. This unpaired electron makes the free radical electrophilic (electron lover), which is highly reactive to attain stability by pairing with nearby molecules even though they may be neutral.

3.4.2 Indirect Action of Radiation

Indirect action occurs if the initial ionizing incidence of low LET radiations takes place on a noncritical predominant molecule (H_2O) and then its products transfer the ionization energy to critical biomolecules (Figure 3.3).

The indirect action of ionizing radiation with cell is the combination of physical, physicochemical, and chemical processes. In general, the water molecule absorbs energy, ionizes, and disassociates into free radicals. This process is referred to as radiolysis of water and it is simply given as follows:

$$H{-}O{-}H \rightarrow H^{+} + OH^{-} \text{ (ionization)}$$

$$H{-}O{-}H \rightarrow H^{\bullet} + OH^{\bullet} \text{ (free radicals)}$$

The possible reactions of radiolysis of water molecules are given as follows. These ions can recombine and form a normal water molecule or they can further produce free radicals and damage the biomolecules.

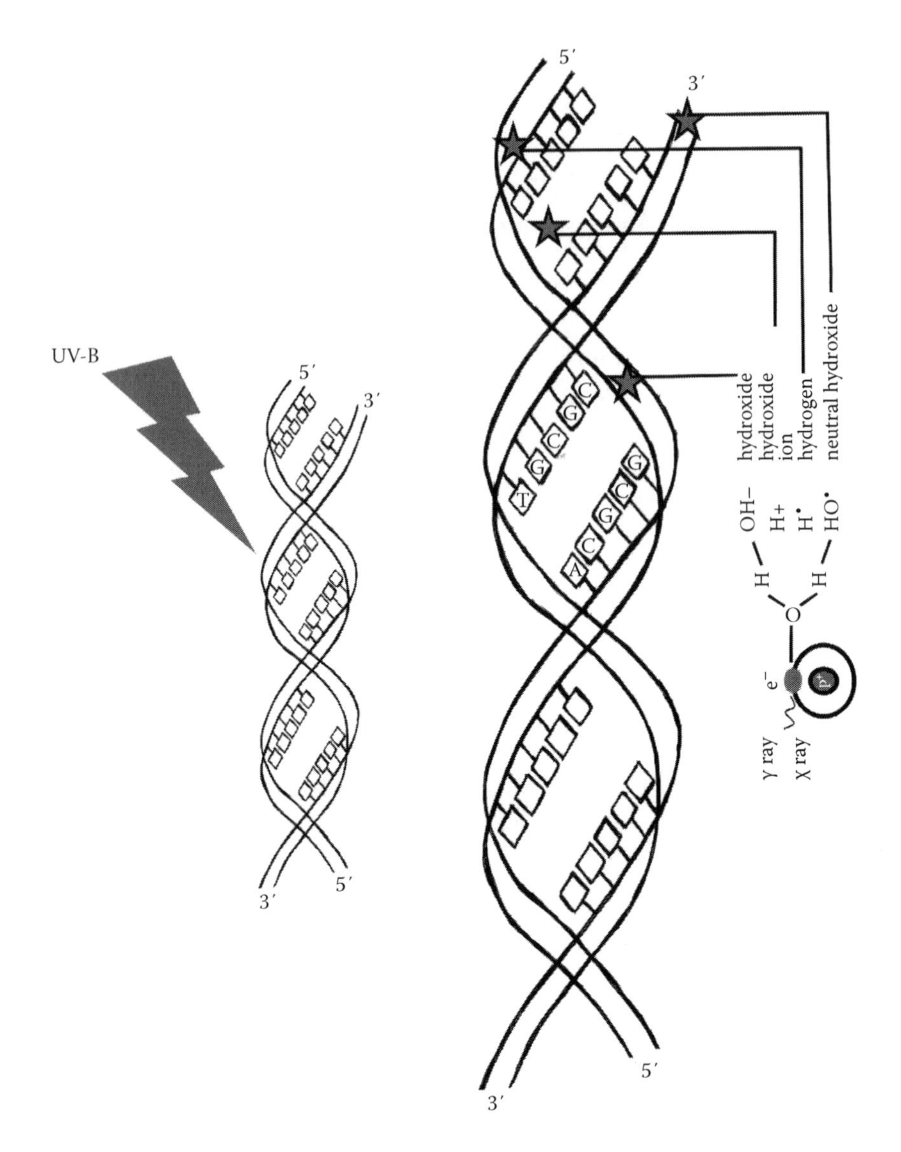

FIGURE 3.3
Direct and indirect action of radiation.

$$H_2O + h\nu \rightarrow H_2O^+ + e^-; \qquad H_2O + h\nu \rightarrow H_2O^* \rightarrow H^\bullet + HO^\bullet$$

$$H_2O + e^- \rightarrow e^-_{aq} \rightarrow H^\bullet + HO^-; \qquad H_2O + e^- \rightarrow H_2O^* \rightarrow H^\bullet + OH^\bullet$$

$$H_2O \rightarrow H^+ + OH^\bullet; \qquad e^- + H^+ \rightarrow H^\bullet$$

$$H_2O^+ + OH^- \rightarrow H_2O + OH^\bullet; \qquad H_2O^+ + H_2O \rightarrow H_3O^+ + OH^\bullet$$

$$H_3O^+ + e^- \rightarrow H_2O + H^\bullet$$

The presence of dissolved oxygen can enhance the reaction by enabling the creation of other free radical species with greater stability and longer lifetimes.

$$H^{\bullet} + O_2 \rightarrow HO_2^{\bullet} \text{ (hydroperoxy free radical)}$$

$$R^{\bullet} + O_2 \rightarrow RO_2^{\bullet} \text{ (organic peroxy free radical)}$$

The transfer of the free radical to a biological molecule causes bond breakage or inactivation of key functions. In addition, the organic peroxy free radical can transfer the radical from one molecule to another causing damage at each step. Thus, a cumulative effect can occur which is more severe than a single ionization effect.

Free radicals readily recombine to make electronic and orbital neutrality. However, when existing in many numbers, orbital neutrality can be achieved by hydrogen radical dimerization (formation of H_2), and the formation of toxic hydrogen peroxide (H_2O_2) occurs as given here. The radical can also be transferred to an organic molecule in the cell.

$$H^{\bullet} + H^{\bullet} \rightarrow H_2$$

$$OH^{\bullet} + OH^{\bullet} \rightarrow H_2O_2$$

$$HO_2^{\bullet} + HO_2^{\bullet} \rightarrow H_2O_2 + O_2$$

The lifetimes of simple free radicals ($H^{\bullet}$ or $OH^{\bullet}$) are very short, on the order of 10^{-5} seconds. Even though they are highly reactive, they do not exist long enough to migrate from the site of formation to the cell nucleus. However, the oxygen-derived species such as hydroperoxy free radical and H_2O_2 are more stable forms, and have a lifetime that is long enough to migrate to the nucleus where they can cause serious damage. It has been estimated that H_2O_2 can cause approximately two-thirds of all radiation damage following the radiolysis of water.

Whether it is direct or indirect action and whatever may be reaction mechanism, all the molecules in the solution finish by returning to an original stable state, but some of them undergo chemical changes. The specific number of molecules (e.g., $OH^{\bullet}$, e^{-}_{aq}, H_2O_2, $H^{\bullet}$, H_2, etc.) formed per 100 eV of absorbed energy for a reaction or a chain of reactions (or 100 eV of energy loss by the charged particle or its secondaries when it stops in water) is known as the radiochemical yield, G. The G value of $OH^{\bullet}$, e^{-}_{aq}, H_2O_2, $H^{\bullet}$ and H_2 is 2.72, 2.63, 0.68, 0.55, and 0.45, respectively (data taken from Duncan and Nias, 1989). As the dose given to the medium increases, the number of these free radicals/molecules also increases significantly. Hence, the concept of radiochemical yield is applied in chemical dosimetry to quantify radiation.

3.5 Interaction of Radiolysis Products with Biomolecules

The short-lived and high-reactive chemical intermediates, in particular free radicals (mainly OH radicals, hydrogen atoms, and hydrated electrons), produced during the indirect action of radiation can easily damage DNA and protein. Electron transfer is a common phenomenon along the chains of DNA and peptides. Electron transfer is often coupled to deprotonation (refers to the removal of proton (H^+) from a molecule).

The two reactions may occur one before the other in a stepwise mechanism, or in a single, concerted step.

Studies show that hydroxyl radical ($OH^•$) is the most chemically reactive of all reactive oxygen species (ROS), with a short lifetime and high rate constant. In vitro studies about the effect of free radicals on biomolecules shows three primary effects on irradiation of biomolecules. They include main-chain scission (breaking of long-chain molecule), cross-linking (formation of a sticky-spur end due to irradiation and its attachment to a nearby natural spur of biomolecules), and point lesions (due to disturbances in a single chemical bond, which leads to late radiation effects like point mutations).

3.5.1 Interaction of Radiolysis Products with Proteins, Carbohydrates, and Lipids

Even though DNA is the primary genetic material, the detailed knowledge of reaction mechanisms in the radiolysis of proteins is also of great importance. The reason is that proteins play a key role in various processes of gene expression and regulation. When radiation interacts with amino acids, products such as ammonia, H_2S, pyruvic acid, CO_2, and hydrogen molecules are produced. In reactions of hydroxyl radicals with peptides, both backbone and side-chain radicals are formed. H-abstraction and N-terminal deamination also occur in peptides. The reactions of hydroxyl radicals at the α-carbon along the protein main chain lead to oxidative degradation to yield amide and keto-acid functions. These reactions extend to chain scission, denaturation, molecular weight modification, changes in solubility, and cross-linking with DNA or with other proteins.

Interaction of radiation/radiolysis products with carbohydrates leads to depolymerization of its monomer molecule (glycogen) and cleavage of α-glycosidic bonds present in glycogen and other molecules. It also oxidizes its terminal alcohols into aldehydes. In addition to these, insulin and blood glucose levels are increased due to the activation of glycogenesis and gluconeogenesis pathways by radiation.

When radiolysis products interact with lipids, the double-bond site and carbonyl groups of these biomolecules are affected. This leads to lipid peroxidation and lipid free radical formation, which are further extended as a chain reaction with other biomolecules to also induce severe cellular damage. As lipids are one of the important constituents of the cellular membrane, disruption of these biomolecules extend to disruption of homeostasis, cellular dysfunction, and finally cell death. In addition to these effects, many important biological functions, namely, digestion, reproduction, neural function, etc., which involve lipids are also modified by irradiation (Cockerham and Shane, 1994).

3.5.2 DNA Damage

Studies with scavenger molecules indicate that almost all of the indirect damage to DNA is due to the attack by the highly reactive hydroxyl radical ($OH^•$). The counterparts of $OH^•$ such as $H^•$ and e^-_{aq} are relatively ineffective, especially for inducing DNA strand breaks.

The water radicals randomly attack all available sites of DNA, but the breaking points of DNA are non-random. Because, the free radical tries to either pass on the extra electron to its nearest neighbor, or grab an electron from the neighbor to make up a pair to revolve around its own nucleus in order to achieve stability. This causes the neighbor to become a free radical, sometimes setting off a chain reaction. By this mechanism,

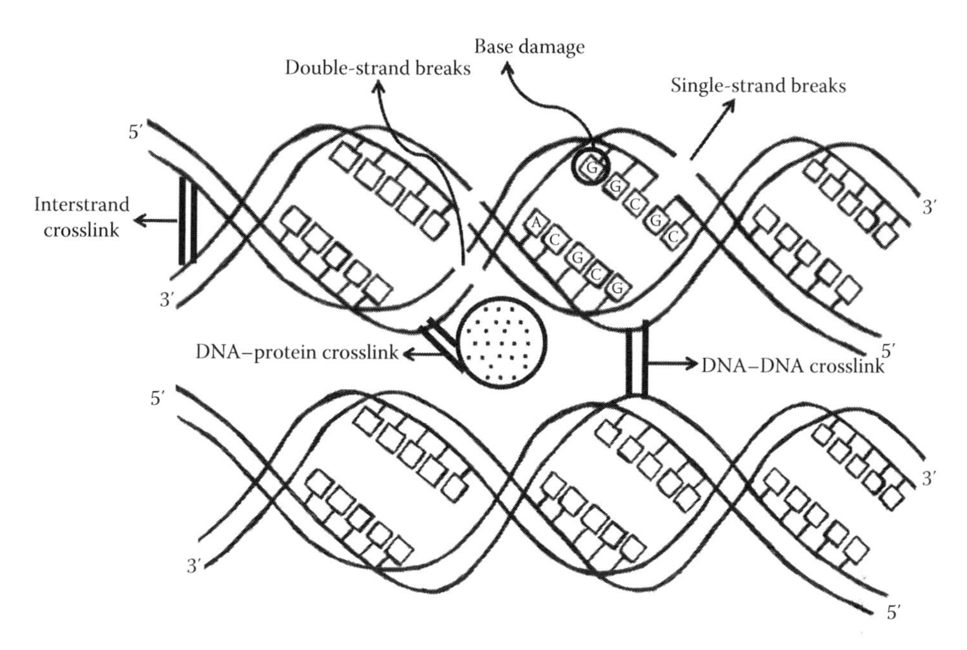

FIGURE 3.4
Various types of DNA damage.

randomly produced stochastic energy deposition events can lead to nonrandom types of damage along DNA at certain distant sites from the sites of the initial energy deposition.

It has been estimated that the water radicals contribute about 60% to the cellular-DNA radiation damage, with different types of lesions, including base and sugar damage, strand denaturation, single-strand breaks (SSBs), double-strand breaks (DSBs), and DNA–DNA and protein–DNA crosslinks (Tubiana et al., 1990). The number of instances of possible DNA damage induced in a cell per 1 Gy of dose is as follows: base damage is 1000–2000; sugar damage is 800–1600; SSB is 600–1200; DSB is 40; DNA–DNA crosslinks is 30, and DNA–protein crosslinks is 150. Out of these, DSB and crosslinks are commonly irreparable (or repairable with error) but others can be repaired. These types of DNA damage are discussed here in slightly more detail and shown in Figure 3.4:

Base damage: Due to the electrophilic nature of OH radicals, a ring double bond is added in the purine and pyrimidine bases. Like the OH radical, the H-atom also acts as an electrophilic and adds C=C double bonds in electron-richer sites. The e^{-}_{aq} molecule induces protonation reactions in the bases.

Hence, the bases can be partially destroyed or chemically modified. Most frequently, they suffer from hydroxylation with the formation of hydroperoxide in the presence of oxygen. The most important of these reactions is the hydroperoxidation of thymine since pyrimidine bases are more radiosensitive than purine bases. They can be represented in the following order of decreasing radiosensitivity (Alpen, 1998):

(most sensitive) thymine > cytosine > adenine > quanine (least sensitive)

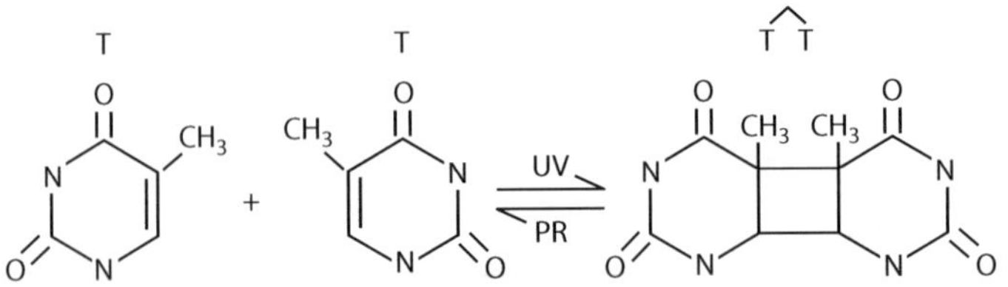

FIGURE 3.5
Formation of the thymine dimer.

For each of these bases, about 20 modifications of the molecular structure have been described and can be identified by chromatography.

Abasic site: Here a base is missing from the DNA but the sugar-phosphate backbone is still intact. This occurs due to a rise in temperature, a drop in pH, or the occurrence of alkylation on the base that causes a break in the N-glycosidic bonds (bond between a base and the sugar-phosphate backbone).

Dimer formation: Dimer formation is one of the important effects of radiation on bases (Figure 3.5). These are formed when two adjacent bases are joined by covalent bonds with the formation of cyclobutane ring between them. Thymine dimers are the most frequent and very stable. They seem to play an important role in the induction of cancers in regions exposed to UV, x-rays, and other chemicals (e.g., Psoralens).

Sugar damage: When the radiation-induced radicals attack the DNA, nearly all H atoms and a major part of the OH radicals add double bonds in the bases. But 10%–20% of the OH radicals react with the deoxyribose sugar by transferring hydrogen. Alternations of deoxyribose are rarer and are not well understood (0.2–0.3 alternations in sugar per 10 SSBs). They can be measured by chromatography. If there is radiation-induced damage, the sugars are oxidized and then hydrolyzed with liberation of the base with or without breakage of the phosphate bonds.

Strand breaks: Strand breaks are the most well-known lesions in DNA. This type of lesion is the most frequent in vitro type of lesion with the highest radiochemical yield. The two types of strand breaks are single-strand break (SSB) and double-strand break (DSB). Experiments with the enzyme endonuclease III, which recognizes a number of oxidized pyrimidines in DNA and converts them to strand breaks, also show evidence for the oxidative base damage formed by the action of OH·. Removal of DNA-bound proteins gives an additional increase in SSBs and DSBs by a factor of 14 and 5, respectively. Owing to the lower molecular weight of the DNA fragments, strand breaks can be studied by comet assay (discussed in Chapter 10). DSBs can be studied in neutral gradients and SSBs by means of an alkaline gradient, which causes the molecule of DNA to open.

Single-strand breaks (SSBs): A single-strand break can take place at the level of the phosphodiester bond (bond between the phosphate and the deoxyribose) or more frequently at the level of the N-glycosidic bond (bond between the base and deoxyribose). A large proportion of the SSB is produced through the action of OH free radicals. Following breakage of the phosphodiester bond, the two strands separate like a "zip-fastener" with penetration of water molecules into the breach and breakage of hydrogen bonds between the bases. There seems to be changes in three or four nucleotides close to the break. SSB are readily repaired using the opposite strand as a template. The numbers of SSBs are three or four times more in well-oxygenated mammalian cells than in hypoxic cells.

The number of SSB is proportional to the dose over a very wide range (0.2–600 Gy). The energy required to produce a SSB is 10–20 eV. An x-ray dose of 1–1.5 Gy produces about 1000 SSB and 50–100 DSB per cell, which is one DSB for every 10–20 SSB. This dose causes the reproductive death of about 50% of mammalian cells. This shows that SSBs are not necessarily lethal and that in a normal cell most of them can be repaired. Specifically, normal cells repair more than 50% of broken molecules correctly, while cells from a patient (with ataxia telangiectasia) can repair less than 10%.

Double-strand breaks (DSBs): A DSB involves breakage of the two strands of DNA at points less than three nucleotides apart. It is homologous if it occurs on the same pair of bases; otherwise, it is heterologous. Heterologous double-strand breaks are more frequent. It can be produced either by a single particle (a localized cluster of ionizations involving an energy transfer of about 300 eV) or by the combination of two SSBs in complementary strands due to two particles traversing the same region before the first break has had time to be repaired. Generally, DSBs are considered to be very dangerous because such breaks are difficult to repair, they can cause mutations, affect genomic stability, and lead to cell death.

The relationship between dose and the number of DSB in mammalian cells is a matter of dispute. Some authors find that the relationship is linear but others it is linear–quadratic. Studies show that the number of DSBs can be correlated with cell lethality in an experimental system over a wide dose range.

Crosslinks: Formation of covalent bonds between bases due to various exogenous (originating externally) or endogenous (originating internally) agents is called DNA–DNA crosslink. This can either occur in the same strand (intrastrand crosslink) or in the opposite strands of the DNA (interstrand crosslink). Crosslinks can also occur between bases (mostly pyrimidine bases) of DNA and polypeptide chain of amino acids. These are referred to as DNA–protein crosslinks. UV light of 254 nm can produce both DNA–DNA and DNA–protein crosslinks. These crosslinks cause replication arrest and cell death if they are not repaired.

3.5.3 DNA Repair

All our human cells are subjected to more than 1 million instances of individual DNA damage per day due to normal metabolic activity, and chemical and radiation exposure. It requires a collection of processes, namely DNA repair mechanism to identify and correct the DNA abnormalities in order to maintain the genome. The rate of DNA repair mechanism depends on the cell type, age of the cell, and the environment. If the repair mechanism fails, the cell will enter into any one of the following four processes:

1. Chromosomal aberration, which may be lethal or nonlethal.
2. Irreversible DNA damage such as DSB and crosslinks.
3. Cell suicide or apoptosis or programmed cell death.
4. Unregulated cell division due to genomic instability, which leads to cancer.

Generally, DNA repair mechanisms are distinguished into the following:

1. Error-free repair mechanisms, which restore the DNA to its original state without any addition or deletion or misrepair. Examples are excision resynthesis, transalkylation, and photorestoration.
2. Repair mechanisms, which are prone to error (misrepair) and so increase the frequency of mutation. An example for this is the SOS gene-regulated repair mechanism.
3. Incomplete repair, which does not reestablish the continuity in the DNA sequence and may lead to cell death.

Once the DNA damage is identified, all the cells undergo an error-free repair process at first and are repaired successfully. If repair of the damage is not possible without error due to very high dose or high LET radiation–induced severe damage, misrepair or incomplete repair may occur.

3.5.3.1 Repair of Base Damage

Base damages are repaired by the base excision repair (BER) mechanism. Primarily, it is responsible for removing small, nonhelix-distorting base lesions (single nucleotide base) from the genome. The BER mechanism is important for removing damaged bases that could otherwise cause mutation by mispair or lead to breaks in DNA during replication. It is performed in the following three basic steps as given in Figure 3.6.

1. The enzyme DNA glycosylase recognizes the incorrect or damaged base and breaks the N-glycosidic bond between the base and dioxyribose sugar. It creates an AP site (either apurimidic or apyrimidic site) in the DNA.
2. An AP endonuclease breaks the phosphodiester bond between the sugar and phosphate near the AP site. The enzyme deoxyribose phosphodiesterase removes the sugar-phosphate lacking the base.
3. By the action of enzyme DNA polymerase I, the correct nucleotide is synthesized and placed. Then, the enzyme ligase links it and joins the gap.

3.5.3.2 Repair of SSB

If one of the strands of a DNA double helix has a defect, the other strand can be used as a template to repair the damaged strand. SSBs are repaired by any one of the repair mechanisms, such as BER, which is same as repair of base damage, nucleotide excision repair (NER), and mismatch repair.

Nucleotide excision repair (NER) mechanism: It repairs damaged DNA, which commonly consists of bulky, helix-distorting damage, such as pyrimidine dimerization caused by UV light. In all organisms, NER involves the following steps (Figure 3.7). To initiate the mechanism, the UvrABC proteins form a dynamic enzyme complex, which recognize damage and remove the damaged part by incision and excision. It comprises three subunits A, B, and C.

1. The subunit A (as the matchmaker) binds with B and forms the AB complex, which moves along the DNA to identify the damaged site by a change in the molecular weight at the damaged site.
2. Once the AB complex reaches the damaged site, the AB–DNA complex is formed, which does not require energy.
3. This is followed by unwinding of DNA (requires energy) so that the subunit A is removed from the site, but the subunit B is strongly associated with the damaged DNA site.
4. This is followed by binding of the subunit C to the B-DNA complex and two incisions (cuts) on both sides of the damaged strand several nucleotides away from the damaged site (usually 24–32 nucleotides in length) are created.
5. The subunit C is released.
6. The damaged part is excised (removed) and the subunit B is released.
7. By the action of enzyme DNA polymerase II, the suitable nucleotides are synthesized and placed. Then, the enzyme ligase links and joins the gap.

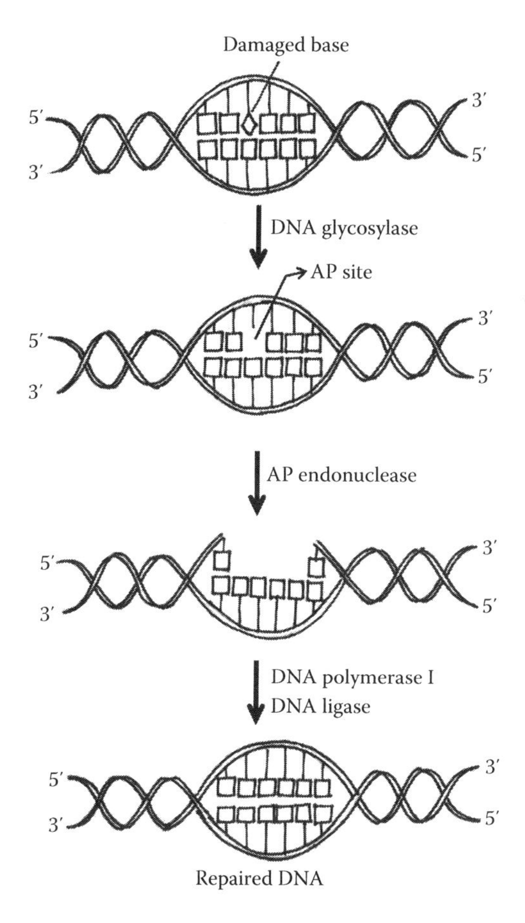

FIGURE 3.6
Schematic representation of BER.

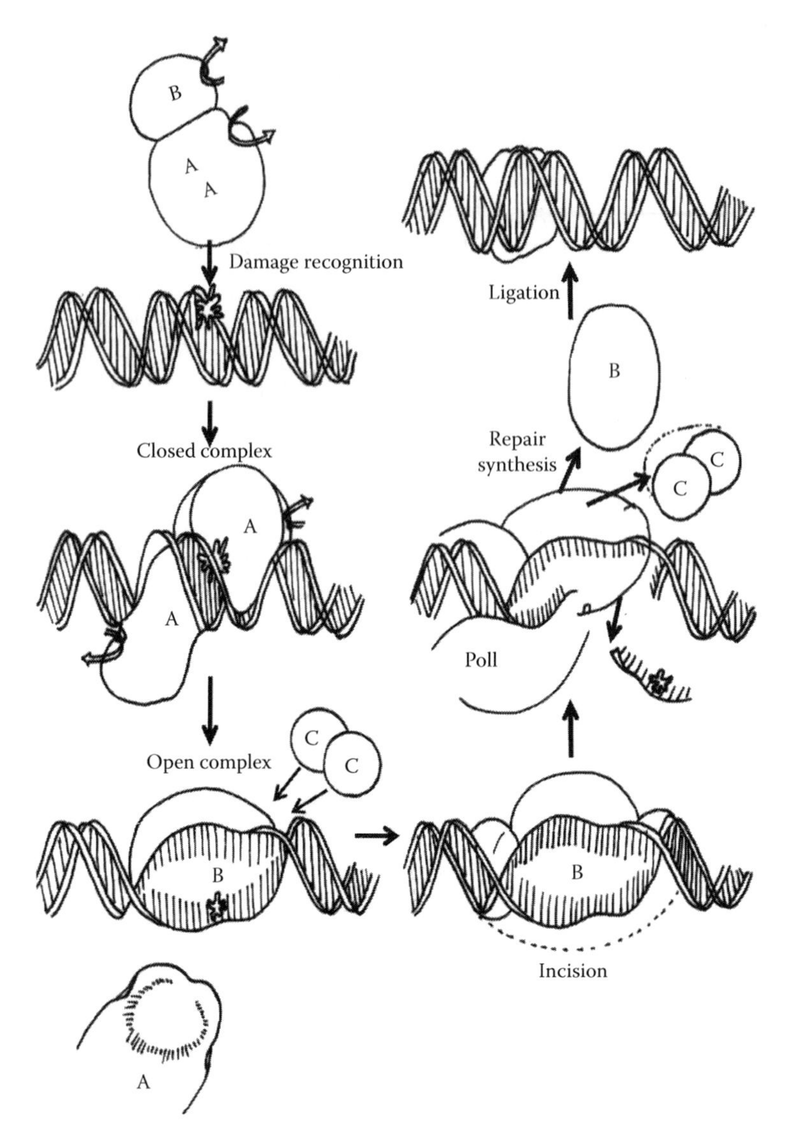

FIGURE 3.7
Schematic representation of NER process in *E. coli*.

3.5.3.3 DNA Mismatch Repair

It is used to identify and rectify the defects in the past repair mechanisms after proof reading. To do so, it uses a number of enzymes as follows (Figure 3.8):

1. The enzymatic protein MutS identifies the mismatch and binds to the DNA.
2. The protein MutL binds at the same position where MutS is binding with the DNA
3. This is followed by cleavage of the single-stranded DNA with error by the action of the enzyme endonuclease (a DNA degradation protein).

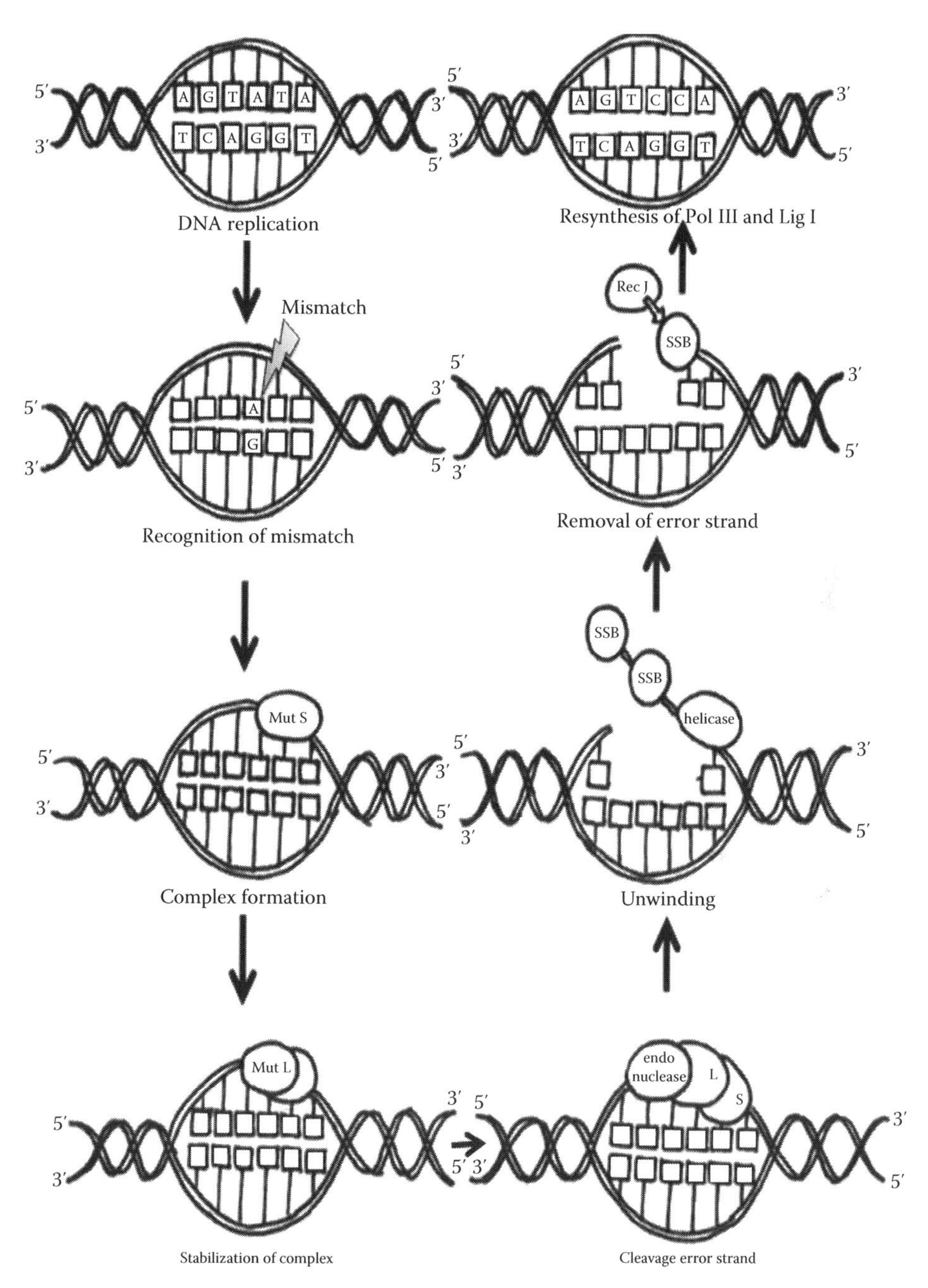

FIGURE 3.8
Schematic representation of DNA mismatch repair.

4. A protein helicase unwinds the DNA and it stabilizes by a single-stranded DNA-binding protein (SSB).
5. SSB binds to the RecJ protein and remove the error strand.
6. Finally, the enzyme DNA polymerase III resynthesizes the correct DNA strand, and the gap is filled by the enzyme ligase I.

3.5.3.4 Repair of DSB

In DSB, both strands of the DNA are damaged, and it may not be repairable without error for all types of radiation because there is no strand with correct gene sequence to use it as a template for repair. Hence, the cell may die in the next mitosis or will be mutated. There are three existing mechanisms to repair DSBs. They are (1) homologous recombination, (2) nonhomologous end joining (NHEJ), and (3) microhomology-mediated end joining (MMEJ).

3.5.3.5 Homologous Recombination

Recombination is induced by the exchange of segments of homologous chromosomes (part of a genome) during meiosis (sexual reproduction) or by transposition of a mobile element from one position to another within a chromosome or between chromosomes. The biological process that uses homologous recombination is used in repair of DSB, recombination with foreign DNA, meiotic recombination, etc. It is a highly accurate and complex process that usually precisely restores the original sequence at the break. In short, the major steps mediated by its set of specialized proteins of homologous recombination (Figure 3.9) are as follows:

1. Alignment of two homologous DNA molecules in the pair of chromosomes and formation of protein–DNA complex at the damaged site
2. Formation of linked intermediates at the damaged site to link two DNA molecules
3. Base pairing between short stretches of the two molecules to form a Holliday junction
4. Migration and resolution of the Holliday junction by DNA cleavage.

3.5.3.6 Nonhomologous End Joining (NHEJ)

It is referred to as "nonhomologous" because the broken ends are directly ligated without the need for a homologous template. It functions in all kinds of cells, and is involved in many other different processes, such as telomere maintenance, insertion of genome, etc. But, it generates small-scale alterations (base pair substitutions, insertions, and deletions) at the break site. Three steps have been identified in the NHEJ process using a series of enzymes. They are (1) DNA end-binding and bridging, (2) terminal end processing, and (3) ligation as given here and shown in Figure 3.9:

1. *DNA end-binding and bridging*: NHEJ is initiated by the recognition and binding of the Ku protein to the broken DNA ends. By forming a bridge between the broken DNA ends, Ku acts to structurally support and align the DNA ends, to protect them from degradation and to prevent promiscuous binding to unbroken DNA. Ku effectively aligns the DNA to allow access to the polymerases, nucleases, and ligases to the broken DNA ends to promote end joining.
2. *Terminal end processing*: This step requires two DNA blunt ends in order to join them together.
3. *Ligation*: Once the blunt ends are in place, the DNA ligase IV place the ligation complex to join the DNA ends together.

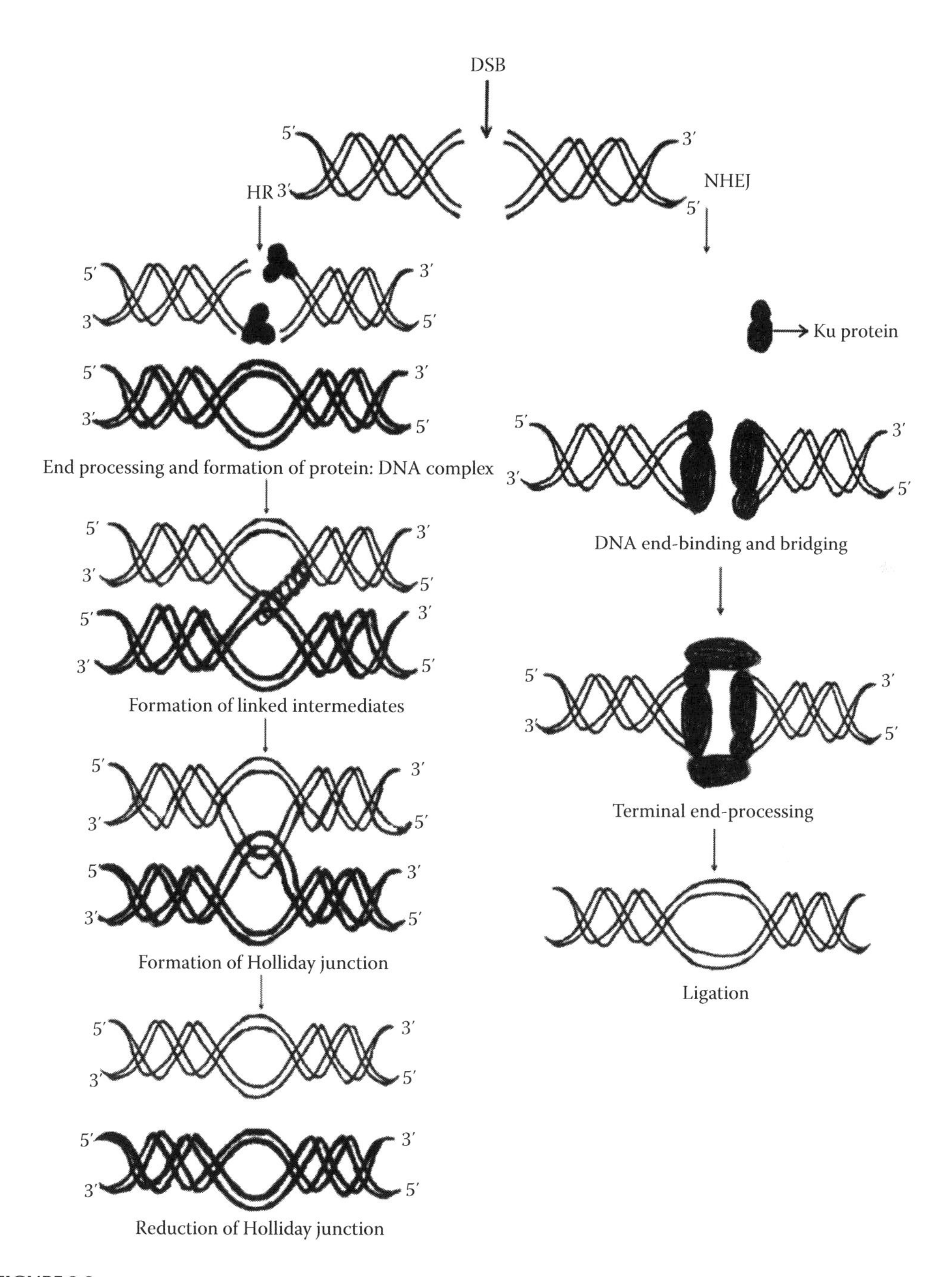

FIGURE 3.9
Diagram of homologous recombination (HR) and nonhomologous end joining (NHEJ).

Microhomology-mediated end joining: Microhomology-mediated end joining (MMEJ) is a major pathway for Ku-independent alternative nonhomologous end joining. It contributes to chromosomal translocations, deletions, inversions, and telomere fusions. It is an error-prone mechanism of DNA repair. The underlying mechanism of MMEJ in mammalian cells is not well understood.

3.5.3.7 Repair of DNA Crosslinks

DNA crosslinks generally cause a loss of overlapping sequence information from the two strands of DNA. Therefore, accurate repair of the damage depends on retrieving the lost information from an undamaged homologous chromosome in the same cell. Retrieval can occur by pairing with a sister chromosome produced during a preceding round of replication. In a diploid cell, retrieval may also occur by pairing with a non-sister homologous chromosome, as occurring during meiosis. Once pairing has occurred, the crosslink can be removed and correct information introduced into the damaged chromosome by the process of homologous recombination repair.

3.5.3.8 Mutation Affecting Repair

As we know, a mutation (deficiency in DNA repair or random nucleotide alterations) is a change in the nucleotide sequence of a short region of a genome. Most of them are point mutations that arise due to the insertion, or deletion, or substitution of single or small number of nucleotides. As it is a part of DNA damage, these abnormalities are also monitored by the repair enzymes present in DNA. These enzymes check the damage in two ways either before DNA replication or after replication in order to repair the mutation at two checkpoints. If there is any malfunction or failure in these two checkpoints, it leads to mutation and vice versa.

3.5.4 Chromosomal and Chromatid Aberrations

Most of the cellular abnormalities due to irradiation are originated from chromosomal aberrations. But, it is not possible to understand the relationship between radiation dose and variation in number of chromosomes to make it as a biodosimeter since this variation is very rare after irradiation. However, it is possible to consider the radiation-induced structural abnormalities (also termed as chromosomal aberrations). There are various types of chromosomal aberrations depending on the position of the cells in the mitotic cycle at the time of irradiation and based on the mechanism of formation.

Depending on the position of the cells in the cell cycle at the time of irradiation, it is classified as chromosomal aberrations and chromatid aberrations. Chromosomal aberrations are produced when the cell is irradiated and not repaired before the cell enters into cell division. Such a kind of aberration induces aberrations in chromatids of both daughter cells. But, chromatid aberrations are produced when cells in the late S phase or G_2 phase of the cell cycle are irradiated, which can damage any one of the two chromatids. An example for chromatid aberration is the "Anaphase Bridge." An anaphase bridge is formed during mitosis when telomeres (ends) of sister chromatids fuse together and fail to completely segregate into their respective daughter cells. Since this event is most prevalent during the anaphase, it is termed as the anaphase bridge. After the formation of individual daughter cells, the DNA bridge connecting homologous chromosomes remains fixed. When the daughter cells reenter into the interphase, the chromatin bridge becomes known as an interphase bridge.

If the cell is irradiated during the S phase and there is partial chromosomal duplication, then both chromosomal and chromatid aberrations are produced. Further, if the cell is in the prophase, sub-chromatid aberrations are produced. This scenario is not same for UV radiation. In the case of circulating blood lymphocytes, the chromosome aberration can be detected easily because these cells are mostly in the G_0 phase of the cell cycle.

Based on the mechanism of formation, the chromosome aberration is classified (Figure 3.10) as (1) deletion (2) intrachromosomal exchange (3) interchromosomal exchange, and (4) agglutination of chromosomes. In general, these are caused by breaking, linking, and exchange of segments.

1. *Deletion or deficiency*: It refers to loss of part of the chromosome. There are two types of deletion. They are (1) terminal deletion and (2) interstitial deletion. Terminal deletion is due to a single break at the end of the chromosome. But, in interstitial deletion, the internal part of the chromosome is lost. This aberration results in an acentric chromosome segment and a chromosome with a centromere, which is a deficient chromosome.
2. *Intrachromosomal exchange*:

   Interstitial deletion: Because of dual cut in between the chromosome ends, a very small chromosome fragment or pair of small chromosome fragments are produced, which look like small coupled spheres.

   Acentric ring: It is also a pair of chromatid fragment in the form of rings without centromeres. As these also appear to be small spheres, it is very difficult to differentiate from interstitial deletion.

   Centric ring: It is also a pair of chromatid fragment in the form of rings but with centromere.

   Inversion: It is a chromosome rearrangement in which a segment of a chromosome is reversed from end to end (e.g., ABAB instead of ABBA). It is possible when a single chromosome undergoes breakage and rearrangement within itself. There are two types of inversions. They are paracentric inversion and pericentric inversion. Paracentric inversions do not include the centromere and both breaks occur in one arm of the chromosome. Nevertheless, pericentric inversions include the centromere and there is a break point in each arm.
3. *Interchromosome exchanges*:

   Reciprocal translocation: In general, reciprocal translocation is termed as the exchange of distal sections (without centromere) of nonhomologous chromosomes. Usually, it is harmless.

   Dicentric or polycentric aberration: These are formed due to the exchange of the proximal part (with a centromere) between two (or more) chromosomes. During this aberration, acentric fragments are also formed. Since the numbers of dicentrics are proportional to radiation dose, it serves as an effective biomarker.
4. *Agglutination of chromosomes*: Due to very high dose of irradiation, chromosomes become thicker and form irregular, viscous, and adherence packets with aggregates of chromatin. This may be due to the structural modification of DNA and proteins, which leads to the formation of chromosomes. Under this situation, the normal cell division is not possible.

3.5.4.1 Stable and Unstable Aberrations

A stable aberration does not affect chromosome segregation so that there is no cell death. It means that these stable aberrations have a single centromere per chromosome and

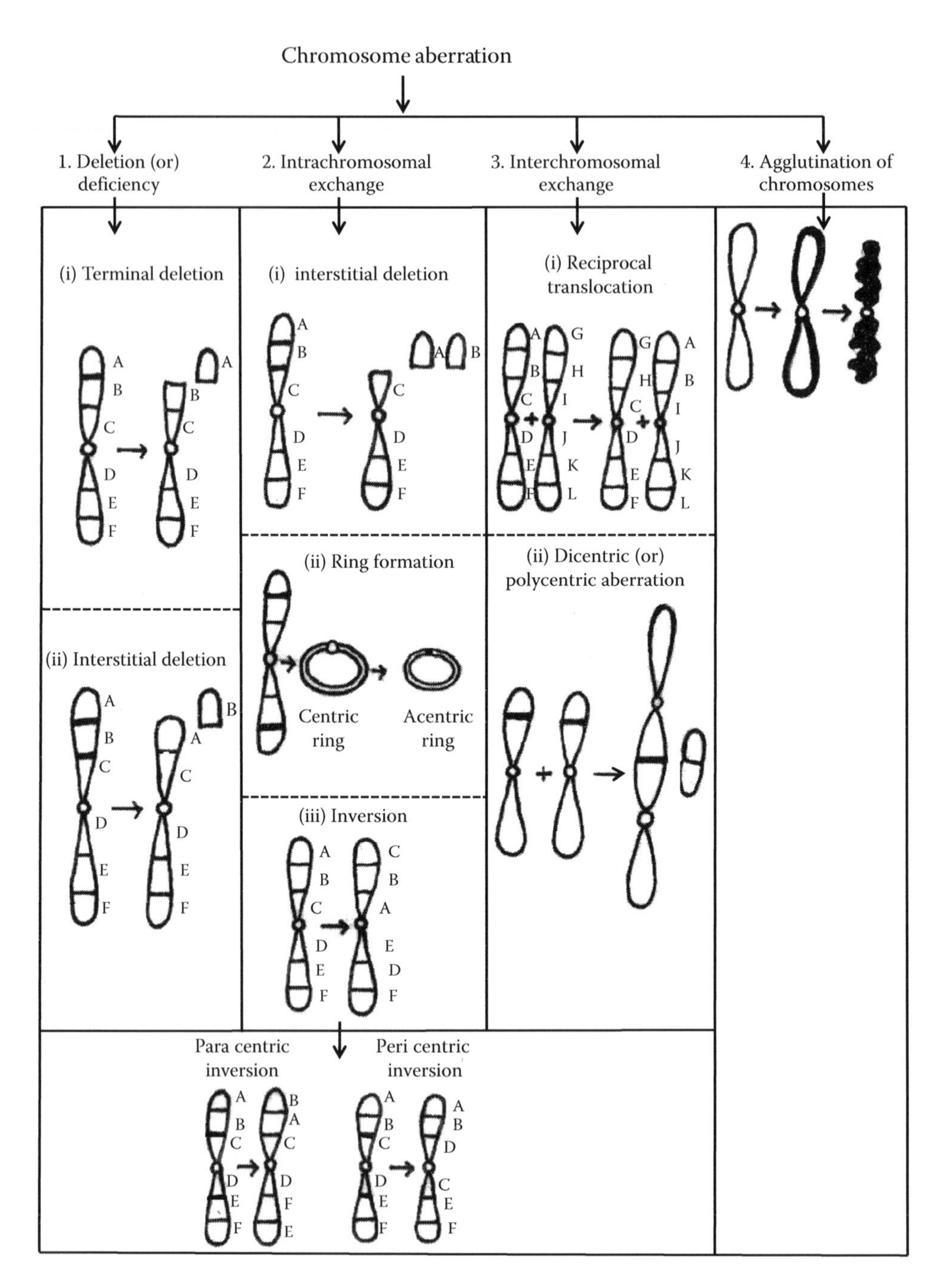

FIGURE 3.10
Various types of chromosomal aberrations.

hence they can divide. So, the altered chromosome persists for many years and can be used as a permanent biomarker to analyze past exposure (e.g., translocation and inversion).

An unstable aberration does not allow chromosomes to segregate properly during mitosis due to the presence of two centromeres in one chromosome and hence leads to cell death. It includes a dicentric ring and an anaphase bridge. It decreases in number over time. Thus, the frequency of unstable aberrations can be applied to analyze recent exposure cases only.

3.5.5 Dose–Response Relationships

The frequency of chromosome aberration increases with the radiation dose to the cells, and hence it serves as a biological dosimeter. *In vitro* irradiation experiments using blood lymphocytes can provide a dose–response relationship that can be used to estimate the unknown dose received by an individual on the basis of the aberration frequency detected in their lymphocytes.

Studies show that the chromosome aberration can be detected at a very low dose even in the range of 0.1 Gy. The number of aberrations and the shape of the dose–response curve depend on the type of aberration. In general, the relation between the number of chromosome aberration and the dose is linear quadratic in the case of dicentrics, reciprocal translocation, centric ring, pericentric inversion, interstitial deletion, and paracentric inversion under low LET irradiation. Figure 3.11 shows the dose–response calibration curve for dicentric yields induced by x-rays. These aberrations require a minimum of two breaks to induce the same, and hence these are decreased if the dose rate is decreased or the fractionation is increased. But, the relation is linear for terminal deletion (fragments), which occurs by a single break.

Studies show that 2 dicentrics per 100 cells are induced at a 0.5 Gy dose of gamma rays and a dose of 2 Gy induces 20 (0.2 dicentrics per cell, i.e., $0.2 * 100 = 20$) (Tubiana et al., 2005). However, for high LET radiation, the dose–response curve shows a linear relationship for all types of aberrations.

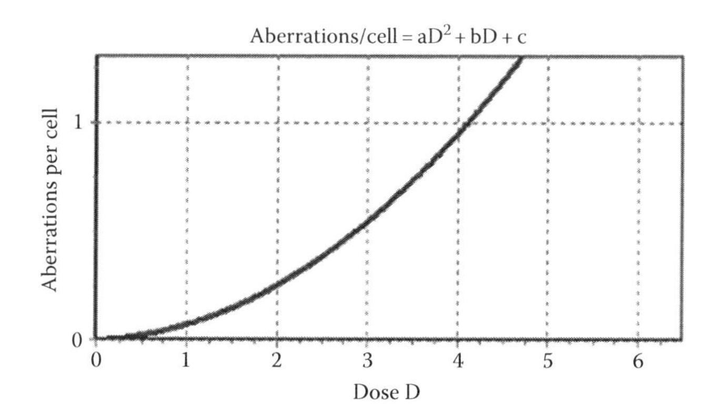

FIGURE 3.11
Dose–response calibration curve for dicentric yields induced by x-rays. (From Lemos-Pinto et al., *Braz. J. Med. Biol. Res.*, 48(10), 2015.)

3.6 Effects of Radiation at the Cellular Level

When the cell is irradiated by ionizing radiation, three major events may occur. They are,

1. *Mitotic delay.* It requires knowledge on the effects of radiation on the cell cycle.
2. *Interphase death.* It is also referred to as premitotic apoptosis. Depending upon the mitotic rate, interphase death occurs. For example, highly mitotic cells undergo interphase death even at lower doses.
3. *Cell death.*

3.6.1 Effects of Radiation on Cell Cycle

As mentioned in Chapter 1 (combine this section with Section 1.7.1), the cell cycle starts at the G_1 phase where a cell gets ready to duplicate itself and then enter into the S phase. In the S phase, it actively copies its genetic material DNA and then enters into the G_2 phase. In the G_2 phase, the cell prepares itself in terms of repairing damage before M phase, mitosis (for somatic cell). After mitosis, the cells can enter the G_1 phase again, or go to the G_0 phase, where they rest.

The sensitivity of cells to radiation varies depending upon the phase of cell cycle in which the cell exists during irradiation. **Sensitivity: M > G_2 > G_1 > S early > S late** (Pawlik and Keyomarsi, 2004). Cells are most radiosensitive in the M phase of the cell cycle by the "law of Bergonie and Tribondeau" (will be discussed in detail later) and most radioresistant in the late S phase due to the possibility of homologous recombination between sister chromatids to repair damage.

Each phase of the cell cycle consists of a checkpoint to temporarily halt the cell cycle. In those checkpoints, such as G_1, S, G_2, and M, the cell takes a decision to continue or not, depending upon the inbuilt genome. DNA damage induced by irradiation can affect the functions of certain proteins (namely, the p53 tumor suppressor), which cause abnormality in the checkpoints.

As shown in Figure 3.12, generally, the G_1 checkpoint monitors the external environment and cell growth, and checks DNA damage. If the DNA is damaged due to irradiation, the G_1 checkpoint can trigger cell cycle arrest, senescence, and even apoptotic cell death, since it is located before the cell cycle commitment point. The S checkpoint also monitors DNA damage. If the DNA synthesis rate is slow it may lead to S phase delay. The G_2 checkpoint monitors the completion of S phase and cell growth and checks for breaks in chromosomes. Then, the M checkpoint is triggered by faulty chromosome segregation, which causes cell death before passing the abnormality onto its progeny. Since the G_1 checkpoint can also arrest cell cycle in addition to delaying it, it is considered to be the most important checkpoint in the cell cycle. But, the other checkpoints, such as S, G_2, and M, can cause cell cycle delay in most of the processes.

The rate of mitotic delay (M phase delay) is influenced by the phase of the cell cycle in which the cells are irradiated. For example, cells irradiated in the G_1 phase show minimal mitotic delay, while cells irradiated in S and G_2 phases show greater delay. Studies show that when there is mitotic delay in the dose range of 0.5–3 Gy, the cells return to near normal later for unknown reasons. At higher doses (>3 Gy), the cells do not recover and the division may not happen, which leads to cell death.

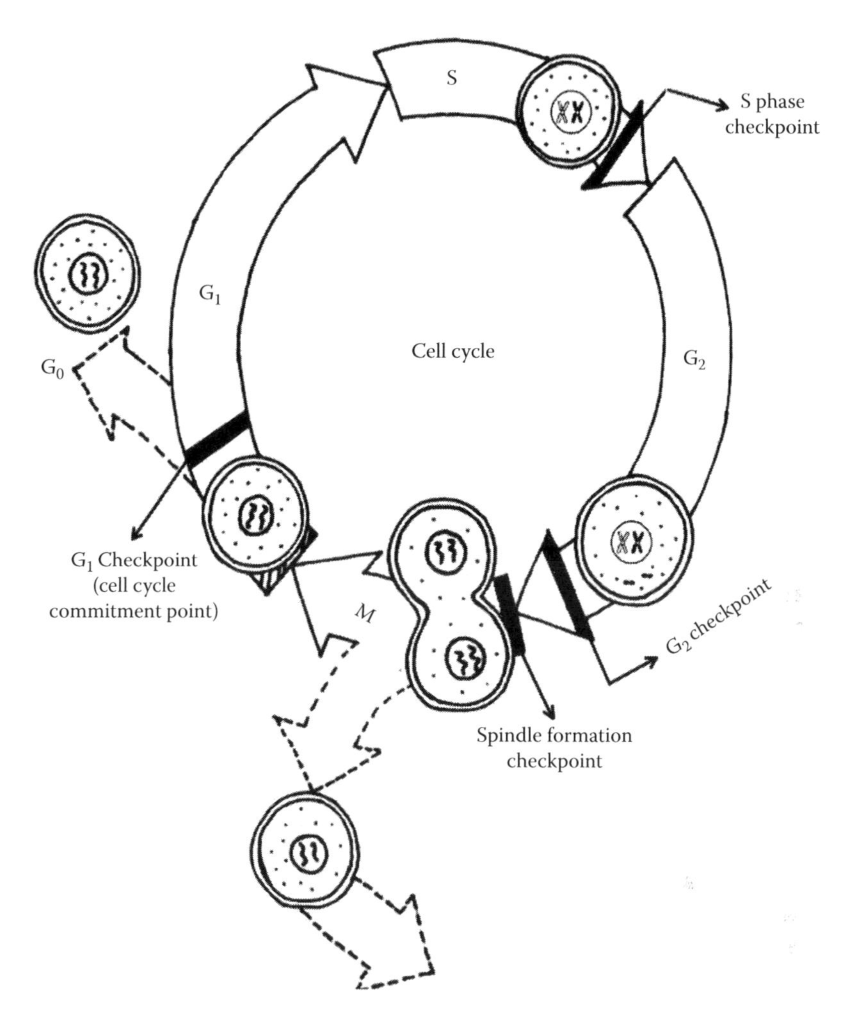

FIGURE 3.12
Various cell cycle checkpoints.

3.6.2 Mechanisms of Cell Death

In general, cell death refers to the loss of ability to divide in proliferating cells (e.g., stem cells, intestinal cells, etc.) or the loss of a specific function in differentiated cells (e.g., nerve cells, muscle cells, secretary cells, etc.). In radiation therapy, radiation achieves its therapeutic effect by inducing different types of cell death (generally, a combination of apoptosis and necrosis) by making changes in the normal genomic stability. Many studies have reported the various genes and intracellular pathways taking part in the different types of radiation-induced cell death. But, still, the exact mechanism that is responsible for radiation-induced cell death is not clear. Hence, the possible and general types of cell death (Joiner and Kogel, 2009) discussed here are as follows:

1. *Apoptosis*—Highly regulated programmed cell death
2. *Autophagy*—To eat oneself

3. *Necrosis*—Accidental death
4. *Senescence*—Death followed by biological aging
5. *Mitotic catastrophe*—Death following aberrant mitosis

3.6.2.1 Apoptosis

Apoptosis is referred to as naturally occurring programmed and targeted cell death. It can be induced by large number of stimuli such as ionizing radiation, UV radiation, death receptors, tumor necrosis factor, growth factor withdrawal, etc. Defects or failure in apoptosis could make cells unable to respond to damage or can develop cancer. Even though there are many studies to explain apoptosis, some related issues, such as intracellular signaling pathways, extension, and its variation in terms of cell sensitivity are not clear. Generally, apoptosis consists of the following steps (Figure 3.13):

1. Cell shrinks and small blebs are formed. Blebs are bulges of the plasma membrane of a cell that are created by the separation of cytoskeleton (microscopic structure of the cytoplasm) from the plasma membrane.
2. The nucleus starts to break apart so that the DNA breaks and fragments into small pieces. During this stage, all the organelles are also located in the blebs.
3. *Cytoskeleton degradation*: On cellular membranes, the cells detach and membranes become more active and fold inward and hence the cytoskeleton of the cell collapses.

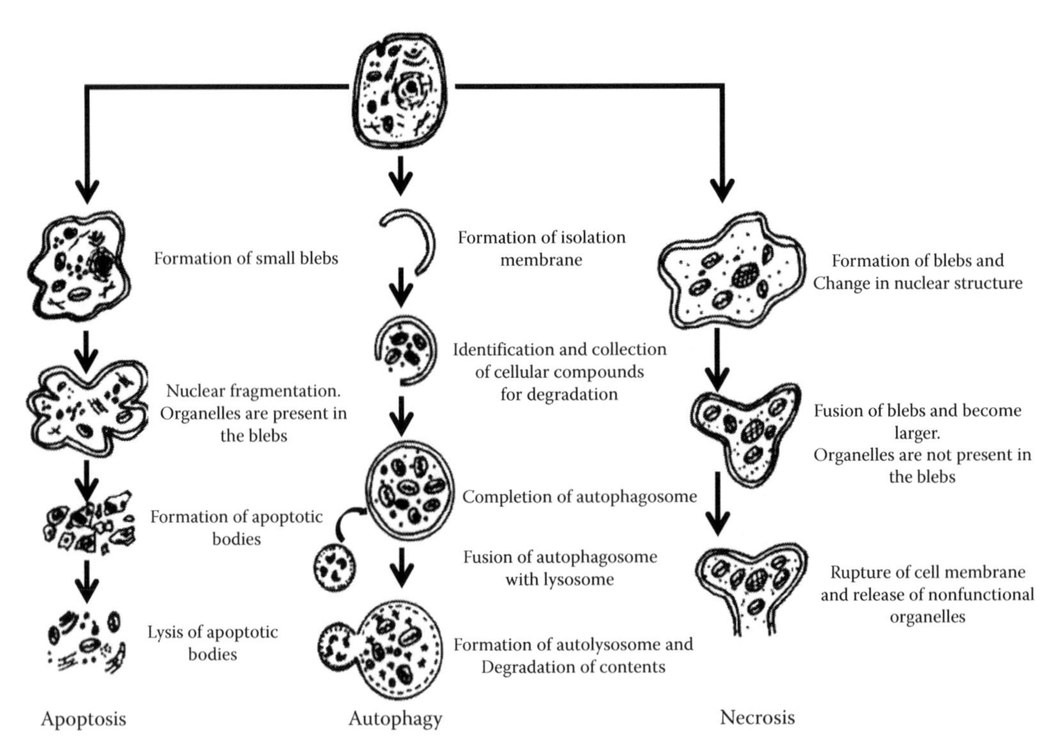

FIGURE 3.13
Schematic representation of apoptosis, autophagy, and necrosis.

4. The nuclear envelope is taken apart into its constituent pieces.
5. These changes convert intact cells into apoptotic bodies, which contain fragmented functional cellular organelles.

Based on the regulatory reactions raised on cell membranes under normal physiological conditions, these apoptotic bodies are recognized and digested by phagocytes (using enzymes in the lysosomes by phagocytosis) without raising inflammatory responses.

3.6.2.2 Autophagy

Autophagy is a more recently described phenomenon. Studies referred it as a genetically regulated form of programmed cell death. It can also act as a cell survival mechanism to synthesize nutrients at the time of starvation in addition to cell death depending on the genetic stimulation. Here, the cell digests itself with the involvement of the autophagic/lysosomal compartment. It is considered to be one type of radiation-induced cancer cell deaths. Since autophagy is also involved in the cell survival, its death mechanism is under controversy. Generally, it consists of the following steps (Figure 3.13):

1. Formation of an isolation membrane
2. Attachment of the cytoplasmic component to a double membraned spherical vesicle (autophagosome) through this membrane
3. Fusion of this autophagosome with the lysosome
4. Degradation of organelles, which leads to the breakdown of the component to eliminate damaged proteins and organelles and also to generate basic nutrients such as amino acids

3.6.2.3 Necrosis

Necrosis refers to an irreversible accidental death of living cells. It occurs when there is not enough blood supply to the tissue either due to irradiation or some other agents such as poisoning. It is executed in the following steps (Figure 3.13):

1. Endoplasmic reticulum and mitochondria are swelled/expanded.
2. Cell lysis occurs, followed by rupture of the plasma membrane, organelles, and nucleus.
3. All the cellular materials, such as degradative enzymes and nonfunctional organelles, are released into the cellular environment, which leads to inflammation and tissue damage.

3.6.2.4 Senescence

Senescence refers to a state of permanent loss of cell proliferative capacity due to cell aging. This can also be induced by radiation-induced oxidative stress, telomere dysfunction/shortening, oncogene activation, etc. These senescence cells are viable but non-dividing, stop to synthesize DNA, become enlarged, and later die mainly by the process of apoptosis. Thus, it is an important mechanism to prevent the growth of cancer cells. In addition to this, senescence also induces cell repair and fuel inflammation associated with aging, which leads to cancer progression as an opposite effect. So, senescence participates in four major processes, such as cell repair, aging, tumor suppression, and tumor promotion. Hence, understanding the benefits of senescence by suppressing its drawbacks is a challenging task.

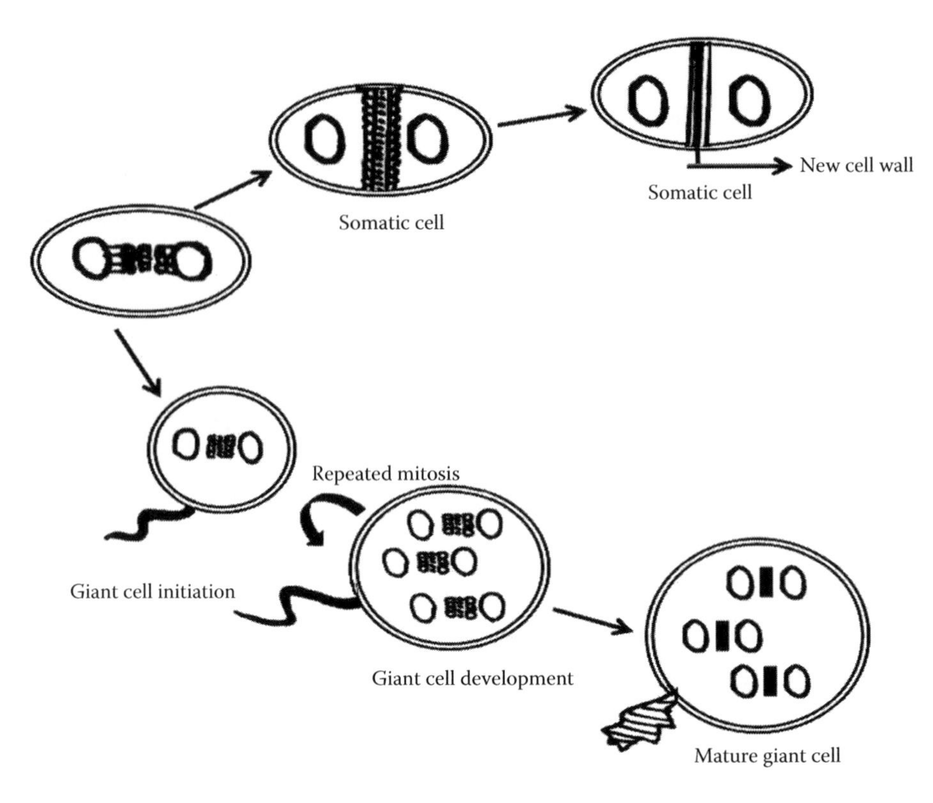

FIGURE 3.14
Difference between normal and giant cells.

3.6.2.5 Mitotic Catastrophe or Mitotic Cell Death

This type of cell death occurs during or after abnormal mitosis and is caused by the formation of giant cells. Giant cells are formed when the cell cycle is arrested in the G_1/G_0 phase due to irradiation. Then, the cells need energy to reenter the cell cycle, so they continue to consume substrates in order to avoid starvation and increase in wet weight and dry weight, due to the continuous synthesis of biomolecules. Further, giant cells also fail to perform cytokinesis (cytoplasmic division) and nuclear division. So, they are not only much larger than the parental cells, but also have an abnormal structure of the nucleus (multiple nuclei) as shown in Figure 3.14 and finally die by apoptosis. The other characteristics of giant cells are cell aging, defects in mitotic assembly, failure of chromosome segregation, duplication of abortive centrosomes, multipolar mitosis, delayed mitosis, formation of monopolar or monoastral spindles, and various microtubular disorders. The giant cell formation as a result of aberrant mitosis is one of the major causes of death in solid tumors.

3.7 Nontargeted Effects of Radiation: Intercellular Communication

It is widely accepted that DNA in the nucleus (DNA is also present outside of the nucleus) is the critical target for radiation-induced cell death and it is believed to be the heart of radiation biology. However, many studies prove that the cell membrane and the mitochondria might also be a target in some instances. It became a challenge to the universally accepted targeted (DNA) effects, and it created a route for the evaluation of non-(DNA)-targeted effects. The essential feature of "non-targeted" effects is that they do not require a direct nuclear irradiation to be expressed, and they are particularly significant at low doses (<0.1 Gy) as well (STUK-A234/DECEMBER 2008).

Hence, a better understanding of nontargeted effects may play a vital role to assess radiation-induced health risk and may give a solution to the existing low-dose dilemma. It is well known that there are many controversies in the assessment of radiation risk at low doses. Many studies have been proposed to show the relation between cancer risk and low-level radiation dose. Unfortunately, none of these approaches has provided clear evidence of cancer induction at low dose levels, but the issue remains highly controversial because of the induction of secondary malignancies in radiotherapy patients.

The most important nontargeted effects of radiation are given briefly in the following. Here, we need to know that the mechanisms associated with these effects are still under investigation.

1. Radiation-induced bystander effects (discussed in Section 4.3.2).
2. *Radiation-induced genomic instability*: It is defined as the appearance of genetic changes in the progeny multiple generations after irradiation of parental cells even though the irradiated parent cell survived and showed no effect.
3. *Radiation-induced adaptive response*: It is a phenomenon that protects cells against subsequent challenging dose by adopting the extracellular signal induced by a small priming dose. Studies show that non-irradiated human lung fibroblast cells were able to adapt if grown in a medium transferred from cells irradiated with either 0.1 Gy of gamma ray or 0.1 Gy of alpha particles. It may be due to the decreased level of p53 protein, increased level of intracellular ROS (reactive oxygen species), and increased level of DNA repair protein AP-endonuclease.
4. *Low-dose hyperradiosensitivity (HRS)*: At very low dose range (below 0.3–0.5 Gy), the radiosensitivity is higher than that predicted from the survival curve model. This may be due to the existence of a very sensitive sub-population of cells, or by the existence of a threshold dose for triggering repair processes. Studies show that this hyperradiosensitivity (HRS) is followed by an increase in radio resistance up to 1 Gy. This is termed as increased radio resistance (IRR).
5. Inverse dose rate effect (discussed in Section 4.2.4).

Objective-Type Questions

1. Which one of the following DNA damage is most important for radiation-induced cell killing?
   a. Base damage
   b. DSB
   c. SSB
   d. DNA–protein crosslink

2. The energy imparted in one event divided by the mean chord length is referred to as
   a. LET
   b. Lineal energy
   c. Specific energy
   d. Stopping power
3. The ionizing radiation that can directly damage the DNA is
   a. Electron
   b. X-rays
   c. Gamma rays
   d. Alpha particles
4. The molecule that has the maximum radiochemical yield is
   a. H_2O_2
   b. $H^\bullet$
   c. $OH^\bullet$
   d. e^-_{aq}
5. The number of possible base damages in a cell induced per 1 Gy of dose is
   a. 1000–2000
   b. 800–1600
   c. 600–1200
   d. 40
6. The following base is the least radiosensitive
   a. Thymine
   b. Cytosine
   c. Adenine
   d. Guanine
7. __________ enzyme is involved in the process of synthesizing new nucleotides.
   a. Endonuclease
   b. DNA phosphodiesterase
   c. DNA polymerase
   d. Ligase
8. _______ is an unstable chromosomal aberration.
   a. Dicentric ring formation
   b. Translocation
   c. Inversion
   d. Substitution

9. The most sensitive phase of the cell cycle is
    a. M phase
    b. G_2 phase
    c. G_1 phase
    d. S phase
10. Highly regulated programmed cell death is
    a. Apoptosis
    b. Autophagy
    c. Necrosis
    d. Senescence

4

Radiation Response Modifiers

Objective

To understand the basics behind the various radiation response modifiers in order to improve clinical outcome and also to think about the need for further research.

4.1 Introduction

The response of normal and cancerous cells to radiation can be modified/influenced by physical, biological, chemical, and technical factors. The physical factors include treatment time, dose, fractionation (TDF) factors, dose rate, volume of tissue irradiated, temperature of the treated volume, and the type of radiation linear energy transfer (LET) used. The biological factors are related to the cell being irradiated, which includes the sensitivity of tissue, bystander effect, and age. The chemical factors are due to the presence or absence of particular molecules, either naturally occurring in the body or administered, which includes radio sensitizers, radio protectors, and radio mitigators. Technical factors are due to inaccuracies in treatment delivery that includes hot spots, insufficient margins, set-up error, geographic miss, etc. The first three factors are discussed here since the technical factor is out of our scope.

4.2 Physical Factors

4.2.1 Treatment Time

Treatment time is one of the important physical factors determining the efficiency of the treatment. It refers to the overall time taken to deliver the prescribed dose from beginning of the treatment until its completion. In conventional treatment, the treatment time is 5–7 weeks in order to get better tumor control with minimal normal tissue complication. Variation in treatment time can modify the tumor control probability (TCP), as well as the normal tissue complication probability (NTCP). For certain cases, the treatment time may be increased or decreased by following the fractionation schedule.

If a longer treatment time is chosen, the tumor cell may proliferate more depending upon the repopulation rate of various kinds of tumors, which will reduce local tumor control. In order to increase tumor control probability, the radiation dose should also be increased especially in rapidly growing tumors (head and neck cancers) to inactivate the repopulated cells. When a shorter treatment time is chosen to treat palliative tumors, then there will be less time for repopulation of early-responding normal tissues (e.g., bone marrow, spleen, etc.), which leads to severe early reactions and, moreover, tumor control will be limited due to significant problems related to hypoxia. But, small lesions and tumors with high therapeutic ratio (e.g., skin tumors) can be treated in a shorter treatment time without much normal tissue complication. However, any shortening in the overall treatment time can decrease the skin tolerance dose to approximately 3–4 Gy/week. Changes in the overall time have lesser effect on late-responding tissues (e.g., skin, heart, etc.) than early-responding tissues since late-responding tissues usually proliferate slowly.

4.2.2 Radiation Dose

The radiation dose (absorbed dose) is defined as the amount of energy absorbed per unit mass of the irradiated medium/tissue. It is used to evaluate any radiotherapy plan in terms of the prescribed dose (dose/fraction and total dose), tumor dose (mean, minimum, and maximum tumor dose), skin dose, etc. From the radiation protection point of view, the terms sublethal dose, supralethal dose, integral dose, etc. are used.

As discussed in Chapter 8, the radiation dose has an impact on the survival curve. It shows that as the dose increases, the number of cells that survive decreases linearly for high LET radiation, but it increases exponentially at low doses and then decreases linearly for low LET radiation.

4.2.3 Fractionation: The Four Rs of Radiobiology

In radiation oncology, it is a well-established fact that multiple radiation doses given over a period of a few weeks give a better curative response than can be achieved with a single dose. During the 1920s, an experiment conducted on exposed sheep testicles found that the differential effect of x-rays on cancer and normal tissues would be better by giving the treatment over a period of 1 week than in 1 day. For example, a total dose of 40 Gy delivered in 20 fractions (5 fractions per week) over a period of 4 weeks with a daily dose of 2 Gy gave a much better outcome (lesser normal tissue damage) than 40 Gy in a single day. It is referred to as the fractionation theory.

Fractionation is defined as dividing the total dose to be delivered into several small portions, allowing a time period between two exposures. It is the very basis for radiation therapy because it enhances the treatment efficacy. Hence, major developments in fractionated radiotherapy have been evolved that relate to the total dose, dose per fraction, interval between fractions, treatment time (see Section 4.2.1), type of tumor, and early- and late-reacting normal tissues. Based on these, dose is optimally distributed into various fractionation schedules, namely hyperfractionation, accelerated fractionation, hypofractionation, split-course treatment, etc. in addition to conventional fractionation as represented in Figure 4.1. For example, in conventional fractionation, a daily dose of 2 Gy is delivered over a period of 7 weeks to deliver the total dose of 70 Gy in 35 fractions. Hyperfractionation is scheduled to spare late-responding normal tissues by increasing the number of fractions, reducing the dose per fractions but by keeping

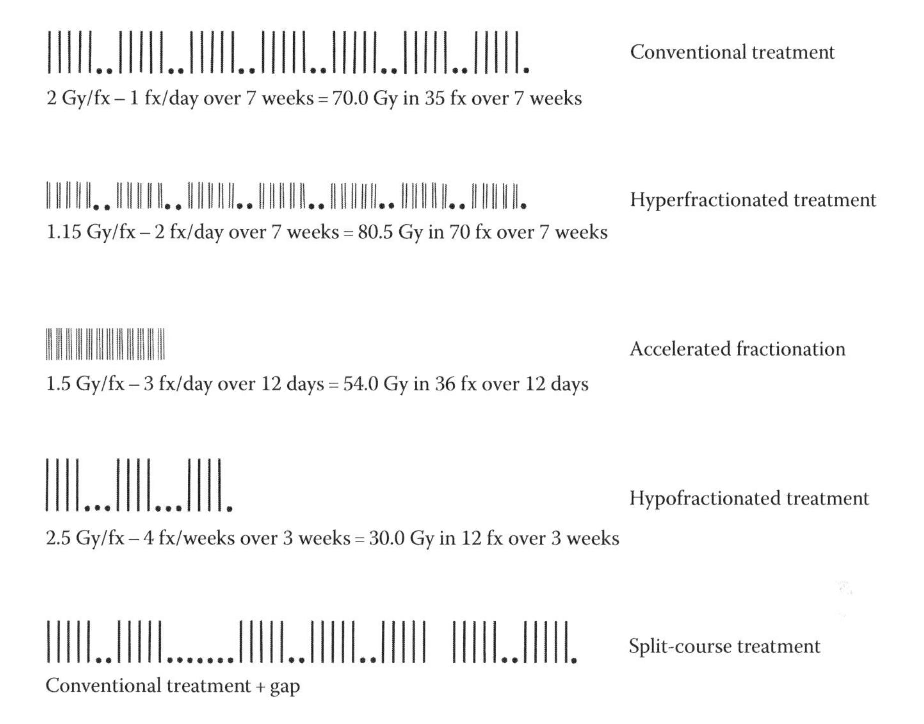

FIGURE 4.1
Various fractionation schemes.

the treatment time the same. Accelerated fractionation is usually scheduled to overcome the accelerated tumor population by reducing the total dose, dose per fraction, total number of fractions, and the overall treatment time. Hypofractionation can be scheduled in busy centers to utilize the machine time well by completing the palliative treatment in a shorter period of time with a higher dose/fraction. Moreover, an efficient scheme is used to treat tumors with a low α/β ratio (e.g., prostate and breast tumors) and also modern techniques, such as stereotactic radiotherapy (cyberknife), use very high dose hypofractionation. In addition, split-course treatment (conventional treatment + gap) allows elderly palliative patients to recover from acute reactions of treatment.

In radiation therapy, many factors influence the selection of appropriate dose/fraction, total dose, and fractionation. They are

1. Radiosensitivity of the tumor: Radiosensitive tumors can be treated by approx. 30 Gy in 4 weeks. But, moderately sensitive tumors require higher doses in the range of 50–60 Gy over a period of 5–6 weeks.
2. Size of the tumor: If the size of the tumor is small, a higher dose can be delivered without compromising normal tissue tolerance.
3. Anatomical location of the tumor. If the tumor is present very close to critical normal structures, only a smaller dose can be delivered depending on the tolerance dose (a tolerance dose is defined as the dose that produces an acceptable probability of treatment complication) of those critical normal structures.

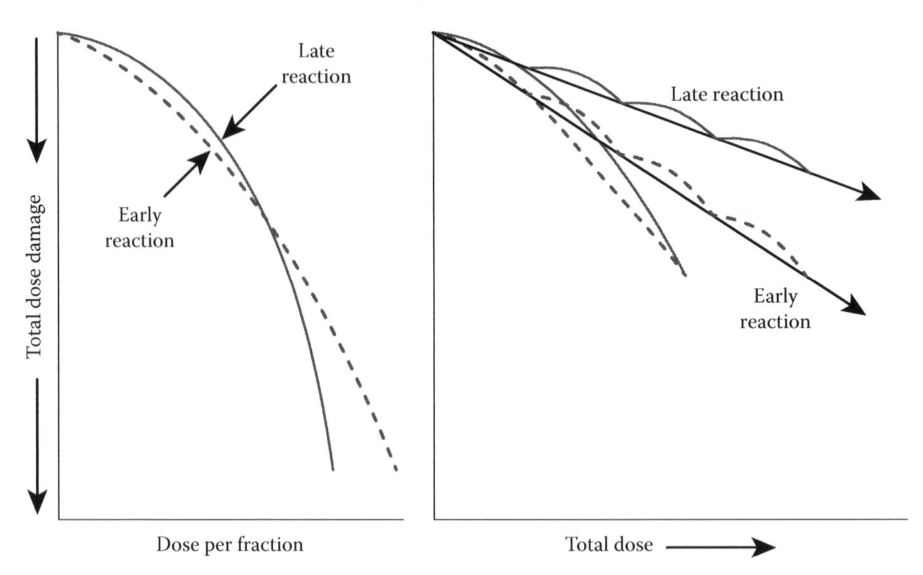

FIGURE 4.2
Difference in cell survival curves for acute and late radiation effects with single or multifractionated doses of irradiation. (Reprinted from *Biologic Basis of Radiotherapy*, Fowler, J.F., Fractionation and therapeutic gain, Steel, G.G., Adams, G.E., and Peckham, M.T. (eds.), Elsevier Science, Amsterdam, the Netherlands, p. 181, Copyright 1993, with permission from Elsevier.)

4. Volume of tissue irradiated (volume effect). This determines the spatial distribution of dose to the normal tissue.
5. Individual patient history like age, etc.
6. Radiation therapy technique (e.g., modern techniques, such as IMRT/VMAT, have the advantage of greater escalation of dose).

Researchers identified many biological factors that can enhance the responses of normal and cancerous cells in fractionated therapy. The most important of them are collectively referred to as "four Rs" of radiation biology. The 4Rs are (1) Repair, (2) Repopulation, (3) Redistribution, and (4) Reoxygenation. One more R, that is, Radiosensitivity is also added, which updates the 4R's into the 5R's of radiation biology. The fractionated treatment has many advantages, which include the following: (1) The acute effects of single dose of radiation can be decreased by allowing the normal cells to repair. (2) Better tumor control can be achieved by redistribution, reoxygenation, and radiosensitivity. But repopulation suppresses the efficiency of fractionation. In order to understand these effects of fractionation, the enhancement in the cell survival curve due to fractionation is shown in Figure 4.2. From this figure, it is observed that the shoulder of the survival curve is repeated many times if the dose is delivered as equal fractions with sufficient time between exposures. Hence, normal cells can survive more than tumor cells due to the combined effect of repair, redistribution, reoxygenation, and radiosensitivity.

4.2.3.1 Repair

It is known that normal cells can repair themselves if they are damaged by radiation. But, in cancer cells, the molecules that decide the repair mechanism themselves are faulty, and hence they cannot be repaired. So repair forms the basis for fractionation. Due to fractionation, normal cells are repaired within a few hours from radiation-induced sublethal damage. Generally, radiation damage to mammalian cells is divided into three types. They are,

1. *Lethal damage*: It is irreversible, irreparable, and leads to cell death.
2. *Sublethal damage (SLD)*: It can be repaired in hours unless additional sublethal damage is caused, for example, by a second dose of radiation. If an additional dose is added, it leads to lethal damage.
3. *Potentially lethal damage (PLD)*: It can be modified by the modification of the post-irradiation environmental condition (e.g., balanced salt solution), which allows the cells to repair themselves, as otherwise, under normal circumstances, it leads to lethal damage.

As discussed in Chapter 3, sublethal damage can be repaired either by a base excision repair mechanism or a nucleotide excision repair mechanism, or a DNA mismatch repair mechanism, or by homologous and non-homologous end-joining.

The initial shoulder on the survival curve after every single radiation dose reflects the ability of cells to repair the radiation damage. If multiple fractions are given with sufficient time intervals (4–24 hours depending on the cell type) between fractions, the effective survival curve becomes straight with a shallower initial slope than the curve for big, single doses. Since different types of tissues have different repair capacities, the interfraction interval and repair half-life is important in conventional fractionated radiotherapy. For example, the spinal cord has a slow repair mechanism with a repair half-life of about 4 hours, and hence it is important to separate the dose fractions by at least 6 hours and preferably 8 hours if two fractions are given on the same day.

4.2.3.2 Redistribution or Reassortment

Generally, cells are distributed in all phases of the cell cycle. As we know (please refer to Chapter 3), cells in the M and G_2 phase of the cell cycle are radiosensitive whereas cells in the S phase are radioresistant. Studies show that fractionated irradiation induces three effects on the cell population. They are (1) blocking cells in the radiosensitive G_2 phase, (2) allowing cells to redistribute and enter into the radiosensitive phase, and (3) recruitment. When these randomly distributed populations of cells are irradiated, the cells in the sensitive phase are killed in the first dose fraction, but the cells in the resistive phase survive. The cells that survive will continue cycling and enter into the sensitive phase (if the interval is smaller than 6 hours) and will be killed by the second or later dose fraction. If the interval is greater than 6 hours then cells will repopulate (Figure 4.3) and degrade the treatment. In some tumors, their stem cells are mostly present in the resting phase (G_0) of the cell cycle, which is a radioresistant phase (since G_0 cells may repair more potentially lethal damage). The fractionation scheme also recruits these cells into the cell cycle in order to reach the sensitive phase and kill the cells efficiently.

4.2.3.3 Repopulation or Regeneration

Generally, repopulation refers to the ability of tissues to replenish themselves following injury. During fractionated radiotherapy, both normal and tumor cells proliferate (growth by the rapid multiplication of cells) faster than before irradiation as there is a lot of cell

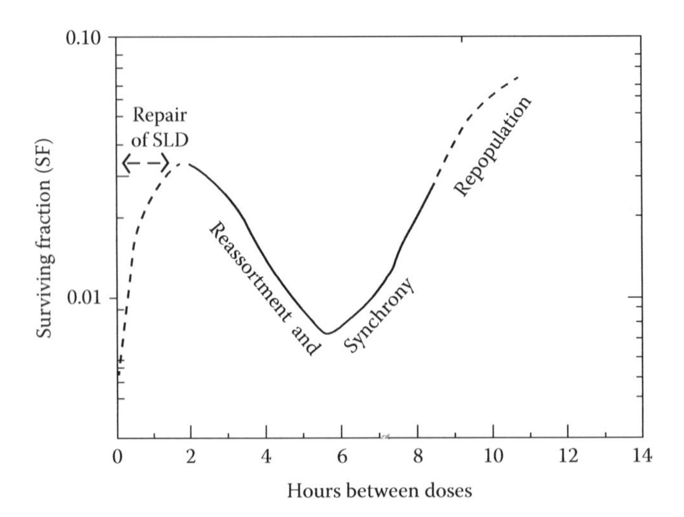

FIGURE 4.3
An example time schedule for repair, reassortment, and repopulation. (With kind permission from Springer Science+Business Media: *Nuclear Medicine Radiation Dosimetry*, 2010, pp. 419, 423, McParland, B.J, Figures 10.16 and 10.21.)

death due to irradiation. This effect is known as repopulation or regeneration. The rate of repopulation is very high and significant toward the end of treatment (3–4 weeks after starting of treatment) because of increased cell death.

Since it increases the number of cancer cells during the course of treatment, it reduces the clinical outcome, and the severity depends upon the types of tissues involved. This effect is predominant in early-responding normal tissues (e.g., skin, mucosa, intestinal epithelium, colon, testis, etc.) and in tumors whose stem cells are capable of rapid proliferation. But, it has little impact on late-responding, slowly proliferating tissues (e.g., spinal cord, bladder, lung, kidney, etc.) and slower-growing tumors such as prostate, breast, etc., which do not suffer very early cell death and early proliferative response. Even though the speed and time-course of repopulation in normal tissues and tumors are poorly understood, the available studies have estimated that local tumor control is reduced by approximately 0.5% each day due to accelerated repopulation in conventional head and neck and cervical cancer patients.

Thus, fractionation must be controlled so as to not allow too much time for accelerated repopulation of tumor cells at the same time, not treating too fast, which may exceed the acute tolerance dose limit. Hence, it is wise to complete the treatment as soon as it starts without a break. But, under unavoidable circumstances, it is better to delay the treatment before starting it rather than introducing a break in the treatment. If not, the tumor volume will be doubled in 4 days due to irregular treatment instead of the normal doubling time of about 2–4 months.

4.2.3.4 Reoxygenation

Generally, tumors consist of aerated (fully oxygenated), hypoxic (low oxygen), and anoxic (no oxygen) cells. When it is exposed to low LET radiation in different fractions, aerated cells are killed in the first dose due to the OER concept (discussed later), hypoxic cells are

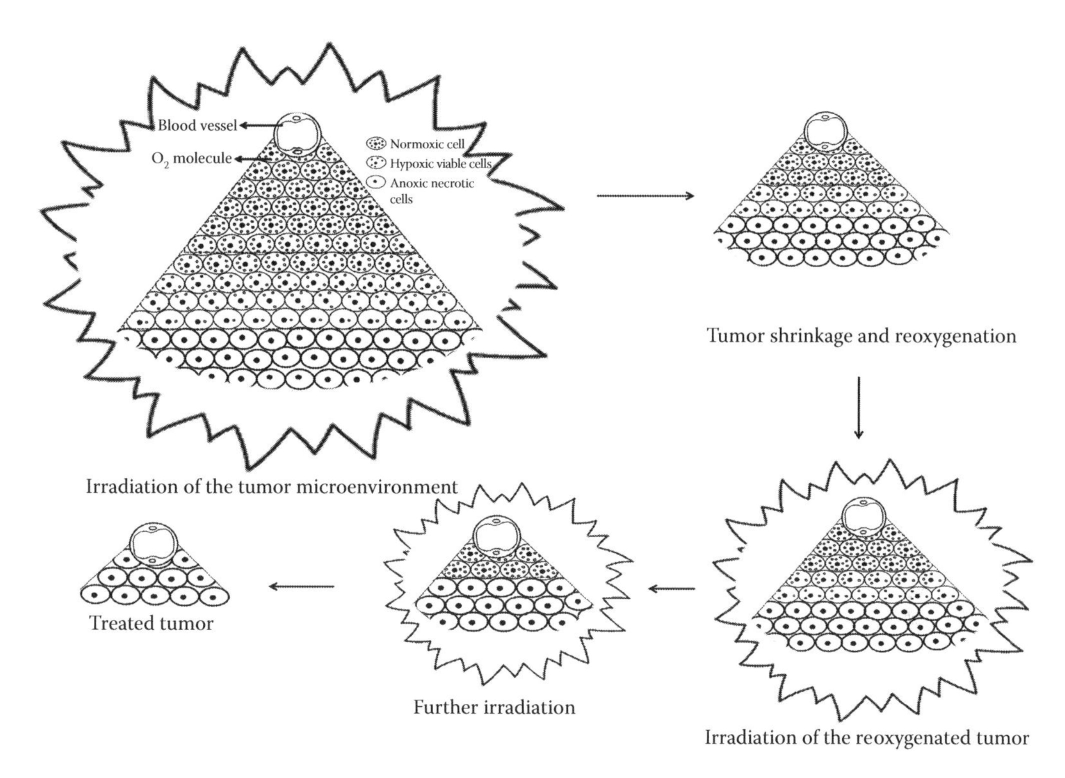

FIGURE 4.4
Schematic representation of tumor reoxygenation.

less damaged, and anoxic cells become necrotic (inactivated). Then, the hypoxic cells are reoxygenated (aerated) either by increased blood supply or redistribution of oxygen through the capillary to the tumor tissue or reduced oxygen utilization by radiation-damaged cells or rapid removal of damaged cells so that the hypoxic cells become closer to functional blood vessels. The reoxygenated hypoxic cells are killed during the next fraction. This process is referred to as "reoxygenation," which is shown in Figure 4.4.

4.2.4 Dose Rate Effect

Dose rate (Dose rate = Dose/time) is defined as the rate at which radiation is delivered to tissues. It is one of the important factors to determine the treatment efficacy. Depending on the dose rate, the type of exposure can be classified as acute or chronic exposure. Acute exposure indicates the delivery of high dose in a few seconds, and chronic exposure indicates the continuous or repeated exposure over months or years. Based on the observations, follow-up, and studies on animals, it is well accepted that acute irradiation is more harmful to the organism than chronic irradiation even though the dose is same because, at a low-dose rate, repairable single-strand break of DNA occurs, but at a high dose rate, irrepairable/misrepairable double-strand break of DNA occurs.

Generally, continuous low-dose rate irradiation may be considered as an infinite number of infinitely small fractions, and hence the survival curve became linear (if the SLD repaired between fractions). Biologically, the differences between low- and high-dose rate irradiation is related to the repair of SLD and redistribution of cells in the cell cycle. It is a

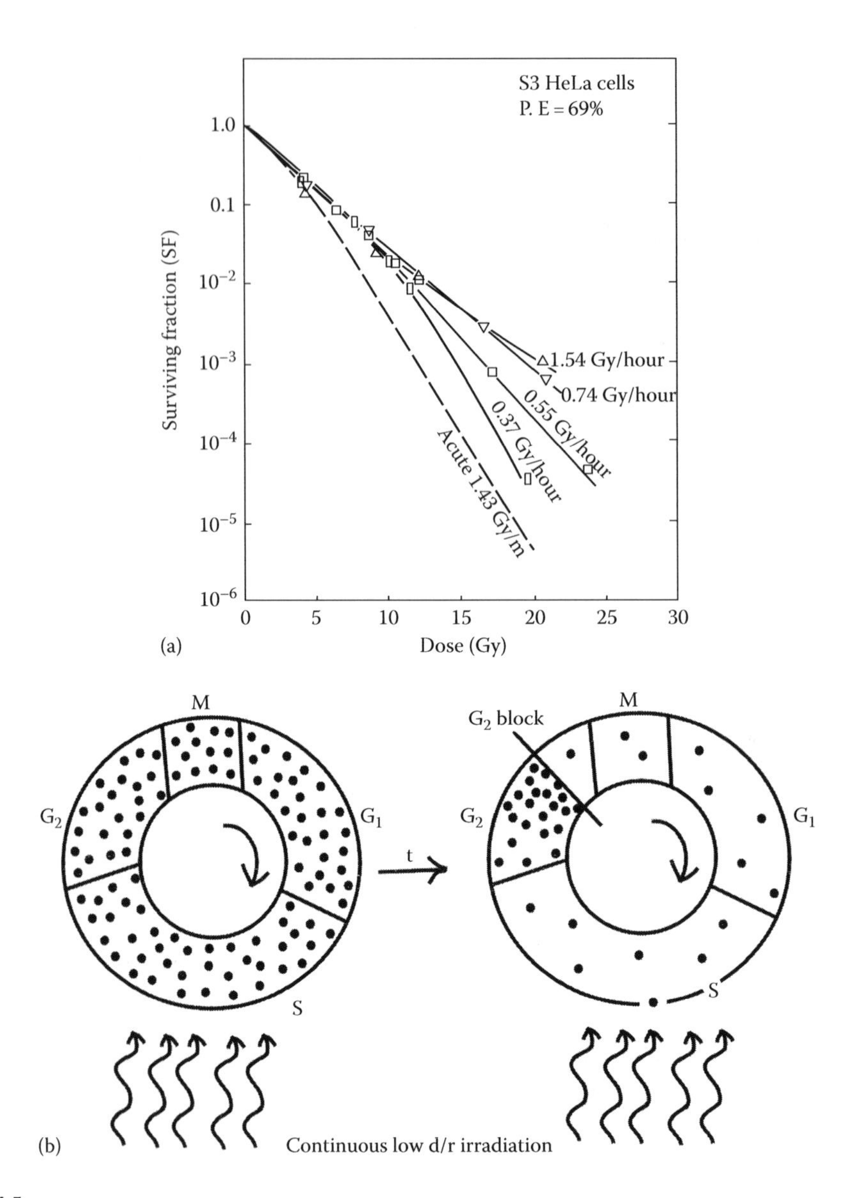

FIGURE 4.5
(a) Graphical representation of inverse dose rate effect, and (b) mechanism of inverse dose rate effect through G_2 block. (Reprinted from *Int. J. Radiat. Oncol. Biol. Phys.*, 21, Hall, E.J. and Brenner, D.J., The dose-rate effect revisited: Radiobiological considerations of importance in radiotherapy, 1403–1414, Copyright 1991, with permission from Elsevier, Figure No: 3.)

general rule that the radiosensitivity of cells (cell killing) is decreased as the dose rate is decreased. In other words, as the dose rate is increased, the cell damage is also increased. This is because of minimal activation of the cell cycle checkpoints at lower dose rates (<0.6 Gy/hour) that repairs sublethal damage and thus minimizes cell killing.

In contrast to the normal dose rate effect, an inverse dose rate effect was reported in HeLa cells when they were exposed to a range of dose rates from 1.53 Gy/hour to 0.37 Gy/hour. At a lower dose rate (0.37 Gy/hour), cells are able to progress through the cell cycle, which blocks the cells in the most radiosensitive G_2/M interphase of the cell cycle. In this way, an asynchronous population becomes a population of radiosensitive G_2 cells (Figure 4.5a and b), which increases the efficiency of cell killing so that the lower dose rate is almost as effective as acute exposure. However, at higher dose rates (>1.53 Gy/hour), cells are "frozen" in various phases of the cell cycle and do not progress further. This effect is referred to as the inverse dose rate effect where the efficiency of cell killing is increased while reducing the dose rate.

4.2.5 Volume of Tissue Irradiated

The radiation response in normal tissues largely depends on the volume of tissue irradiated in addition to the number of cells killed. This is referred to as the "volume effect." It plays an important role in analyzing the risk factor involved in any treatment. Studies show that the volume effect is varied among the various organs of the body depending on their structural organization. All the organs are structured as the collection of many number of functional subunits (FSU). Based on the arrangement of FSUs, organs can be classified as (a) serial organs (linear or tubular structural organization), for example, spinal cord; (b) parallel organs, for example, kidneys, lungs, liver; (c) serial–parallel organs, for example, heart; and (d) a combination of parallel and serial structures, for example, a nephron (ICRU Report 62, 1999). The illustration of all these structural organizations is shown in Figure 4.6.

A FSU is an organelle tissue volume that can be regenerated by a single stem cell. Studies expressed that organ function depends upon the retention and aggregation (combined effect) of stem cells in these individual FSUs. If the organ consists of at least one undamaged stem cell in all the FSUs after irradiation, then irradiation would not affect the reserve capacity of the irradiated organelle volume to sustain their normal function. But, the damaged FSU (no undamaged stem cells) cannot be rescued even by migration of stem cells from nearby FSUs.

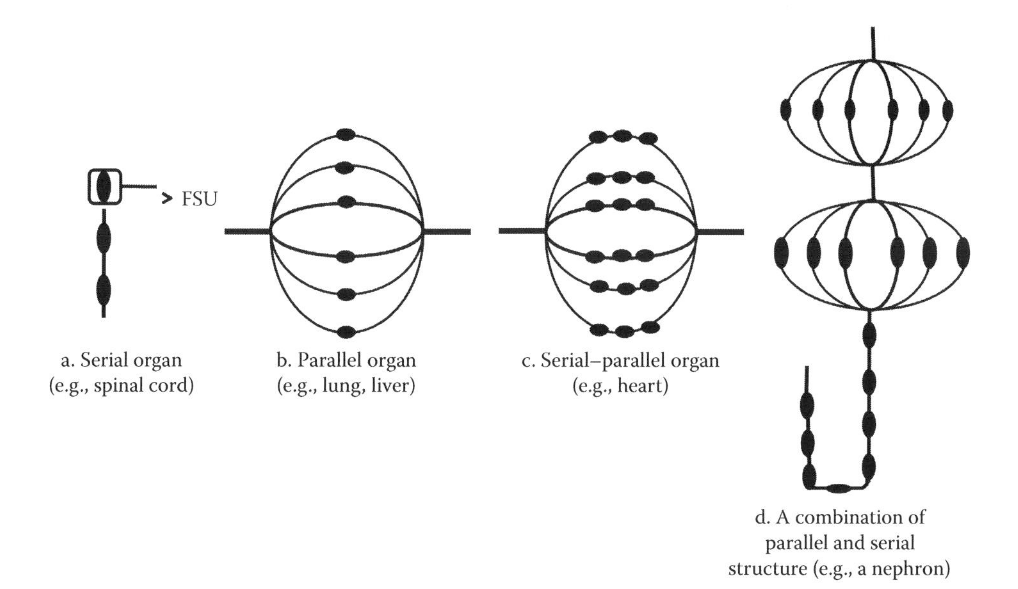

FIGURE 4.6
An illustration of various structural organizations. (From Donald, C.J. and Nahum, A.E., *Radiotherapy Treatment Planning: Linear-Quadratic Radiobiology*, CRC Press/Taylor & Francis, Boca Raton, FL, 2015, p. 113, Figure 9.6. With permission.)

Due to the serial structural organization of spinal cord, a high dose to its very small volume can cause paralysis (large volume effect) even though it has relatively radioresistant tissues and also the dose to the remaining volume is negligible. Since the volume effect is large, the tolerance dose increases steeply with decreasing volume. However, late responding organs such as kidneys, lungs, and liver have parallel organization of their FSUs and hence they can withstand their function even though the major part of their volume was exposed to high dose and radiosensitive too.

4.2.6 Temperature

The amount of heat energy required to induce cell death is 1000 times greater than radiation energy. Reproductive cell death due to DNA damage is less common in heat. But heat induces cell death by damaging cellular plasma and cytoskeleton membranes. Unlike radiation-induced cell death, heat produces immediate cell death, and cells in the S phase are most heat sensitive. Of course, as the temperature of the treated volume increases, cell death is also increased. For example, at temperatures between 40°C and 43°C there is minimal cytotoxicity, but it induces cell killing at higher values (>45°C). This is referred to as hyperthermia. It is used in combination with radiation for the treatment of superficial malignancies as a radiosensitizer since it enhances radiation-induced DNA damage and reduces repair of SLD. This is represented in terms of "Thermal Enhancement Ratio." It is defined as the ratio between x-ray doses without heat to the x-ray dose with heat to induce the same biological effect. Even though heat enhances irradiation efficiency, it has limited application in combination with radiation therapy due to thermo-tolerance induction.

4.2.7 Linear Energy Transfer and Relative Biological Effectiveness

Linear Energy Transfer (LET) or restricted linear collision stopping power describes the mean energy locally imparted to the medium by a charged particle of specified energy when it travels unit distance, that is, the amount of energy imparted/lost/transferred per unit path length LET = dE_{Δ}/dl. It is represented in the unit of keV/μm or MeV/cm. It is varied depending upon the type of radiation as the particles have different mass and charge. LET is inversely proportional to the square of the velocity (related to mass) and directly proportional to the square of the charge. It is represented as LET α $Charge^2/Velocity^2$. Based on LET, ionizing radiation can be classified as follows:

1. Low LET radiation (<10 keV/μm), for example, x-rays and gamma rays, electrons, and protons.
2. Medium LET radiation (10–100 keV/μm), for example, neutrons.
3. High LET (>100 keV/μm) radiation, for example, alpha particles and heavily charged particles.

Electromagnetic radiations such as x-rays and gamma radiations have no mass or charge. They interact with matter by producing fast electrons that have a small mass and −1 charge. Because of the electron's high speed and low mass the interactions that are produced are far apart from each other, and for this reason electromagnetic radiation is called low LET radiation. Hence, the energy of low LET is spread throughout the cell and produces a more uniform distribution throughout the cell, which causes lesser cell damage (more cell survival).

Neutrons, even though they have no charge, are highly ionizing particles because of their mass. They do not interact with orbital electrons but they interact with the atomic nuclei and produce charged particles (mainly protons). Alpha particles and heavily charged particles (carbon ion, Fe ions, etc.), because of their mass and charge, are even more highly ionizing than neutrons. These radiations produce well-defined dense tracks in a short distance that cause severe cell damage along their path (lesser cell survival). The typical LET values of different types of radiations are given in Table 4.1 and the variation in cell survival for three different types of radiations (low, medium, and high LET) are shown in Figure 4.7.

TABLE 4.1

Typical LET Values of Different Types of Ionizing Radiations

Type of Radiation	LET Values (keV/μm)
Co-60 gamma ray	0.2–0.3
250 kVp x-rays	2.0
3 MeV electrons	0.2
Protons	0.5–5.0
Neutrons	12–100
Alpha particles	100–150
Heavy charged particles	100–2500

Source: Data taken from Tod, W.S., *Targeted Radionuclide Therapy*, Lippincott Williams & Wilkins, 2011.

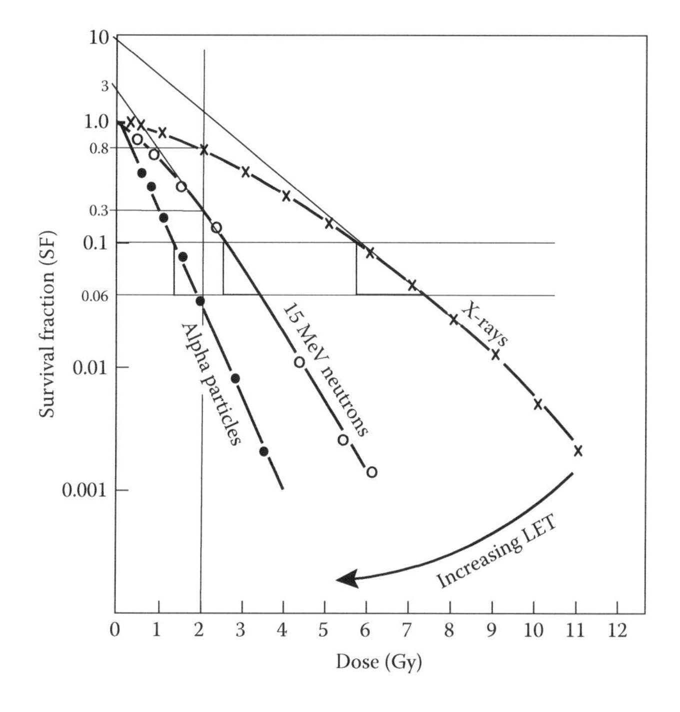

FIGURE 4.7
Variation in cell survival for different types of radiations. (From Kelsey, C.A., Heintz, P.H., Sandoval, D.J., Chambers, G.D., Adolphi, N.L., and Paffett, K.S.: *Radiation Biology of Medical Imaging*, 1st edn. p. 95. 2014. Figure 5.9. Copyright Wiley-VCH Verlag GmbH & Co. KGaA. Reproduced with permission.)

From Figure 4.7, the following information is observed:

1. As the LET increases, the slope of the cell survival curve increases (becomes steeper) and D_0 (0.1/0.037) becomes smaller.
2. As the LET increases, the extrapolation number (n) goes to one (the shoulder becomes smaller and Dq becomes smaller). The value of n for alpha particles, neutrons, and x-rays is 1, 3, and 10 respectively. The value of Dq for alpha particles, neutrons, and x-rays is 0, approx. 60, and 240 cGy, respectively.
3. The biological effects produced by three different LET radiations are different even at same dose. For example, the survival fraction (SF) of cells irradiated at 200 cGy by x-rays, neutrons, and alpha particles is approx. 0.8, 0.3, and 0.06, respectively. It may be further explained in terms of relative biologic effectiveness (RBE).

The biological effect produced by the same doses of different LET radiations is not same. Hence, the term RBE, which relates the ability of standard radiations to different LETs to produce a specific biological effect, is introduced. RBE is a comparison of a dose of standard 250 kVp x-rays to a dose of various test radiations that produce the same biologic response. RBE is expressed as follows:

$$\text{RBE} = \frac{\text{Dose}\left(\text{Gy}\right)\text{ from standard 250 kVp x-rays}}{\text{Dose}\left(\text{Gy}\right)\text{ from test radiation to produce the same biological effect}}$$

RBE varies with many factors such as the radiation type (low, medium, or high LET), type of cell or tissue (sensitivity and age), biological environment (oxygenated or aerobatic), biological endpoint (SF = 0.1 or 0.001), dose, fractionation, and dose rate. As the dose is decreased, the RBE is generally increased. The RBE for a fractionated treatment is greater than for a single exposure, because a fractionated schedule consists of a number of small doses and the RBE is larger for small doses. As the survival is higher at the lower dose rate, the RBE is also varied with dose rate. In order to further understand RBE, the survival curves of neutrons and x-rays are shown in Figure 4.8 and discussed as follows:

$$\text{We know,}\quad \text{RBE} = \frac{\text{Dose}\left(\text{standard x-rays}\right)}{\text{Dose}\left(\text{test neutrons}\right)}$$

$$\text{From Figure 4.8,}\quad \text{for SF of 0.1,}\quad \text{RBE} = \frac{5.8\text{ Gy}}{2.6\text{ Gy}} = 2.2$$

$$\text{for SF of 0.001,}\quad \text{RBE} = \frac{11.75\text{ Gy}}{6.3\text{ Gy}} = 1.9$$

It shows that as the biological endpoint (SF) is changed, the RBE is also changed, which is expressed in Figure 4.9 for different LET radiations as well.

From Figure 4.9, the following information is observed:

1. For diagnostic x-rays, the RBE is approximately 1 as it is considered to be a reference radiation.
2. At lower LET (<100 keV/μm), the RBE is also low. Because the number of hits (refers to damage to the sensitive targets) is low per particle, the dose required to

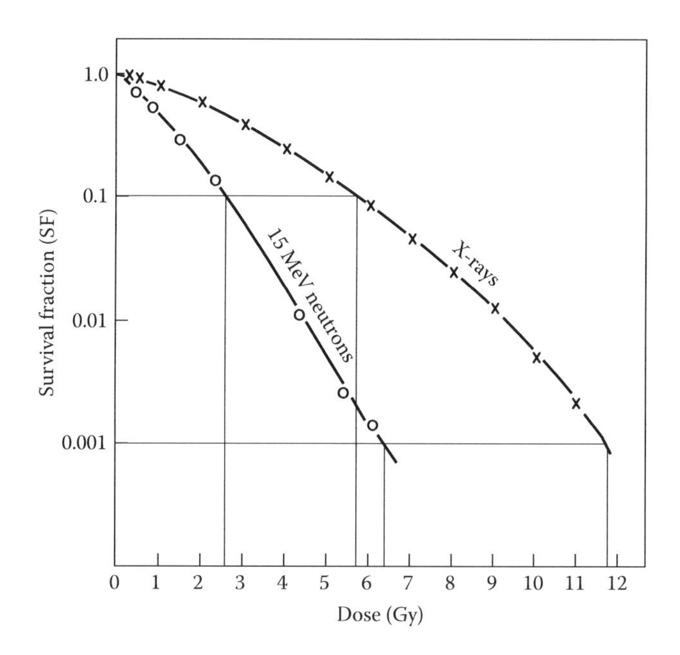

FIGURE 4.8
Survival curve of neutrons and x-rays to express RBE. (From Kelsey, C.A., Heintz, P.H., Sandoval, D.J., Chambers, G.D., Adolphi, N.L., and Paffett, K.S.: *Radiation Biology of Medical Imaging*, 1st edn. p. 95. 2014. Figure 5.9. Copyright Wiley-VCH Verlag GmbH & Co. KGaA. Reproduced with permission.)

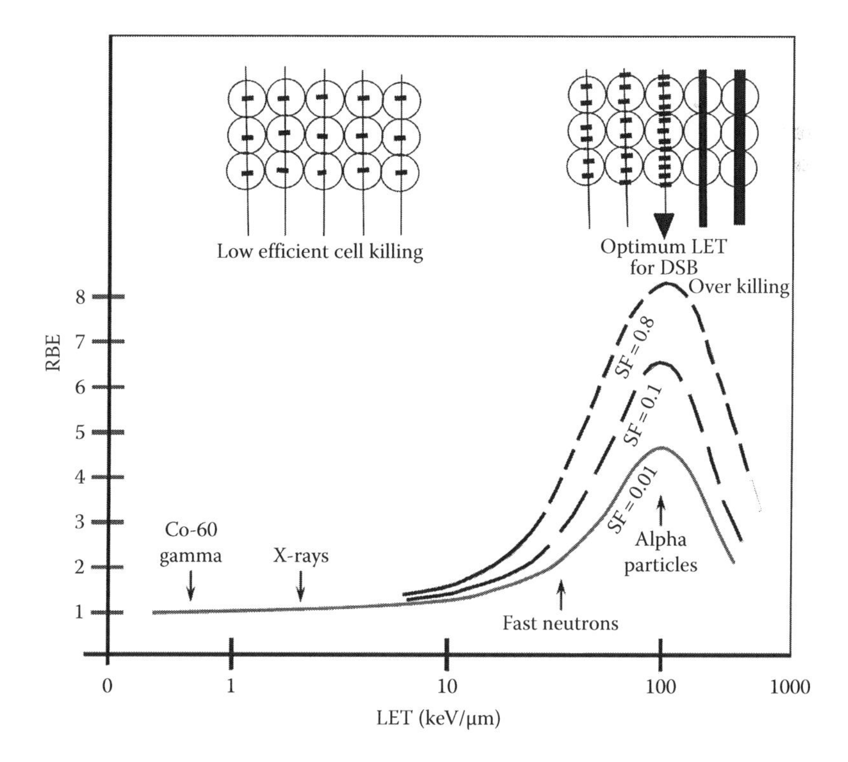

FIGURE 4.9
Variation in RBE in terms of LET and biological endpoint (SF).

kill the cell is higher as compared to higher LET radiation. As the LET (from low to medium) and SF (from 0.01 to 0.8) increase, the RBE also increases proportionally.

3. RBE reaches a maximum at about 100 keV/μm (the optimal LET), which is independent of the endpoint (SF) used.
4. At optimal LET (=100 keV/μm), the RBE is maximum. This means that the number of hits per particle is just at the right level to kill the cell with no dose wasted.
5. At higher LET (>100 keV/μm), the RBE is reduced further due to the "overkill effect" and it reaches unity at about 200 keV/μm because the number of hits per particle is higher than the number required to kill a cell. Hence, the dose deposited by these excess hits will be a wasted dose because the dead cells will not die once again with more hits.

4.3 Biological Factors

4.3.1 Radiosensitivity of Tissues

It is well known that the response of the tumor or tissues to irradiation basically depends on the radiosensitivity of the individual cells. Radiosensitivity refers to the degree of shrinkage of a tumor/normal tissues following irradiation and it is expressed based on the law of Bergonie and Tribondeau (1906) as given here:

1. High proliferation (reproduction) rate and fast growth rate increase the radio sensitivity of the cells.
2. Undifferentiated cells (immature cells) or cells in the process of differentiation (young cells) or cells undergoing active mitosis are highly radiosensitive, for example, stem cells.
3. Cells have a high metabolic rate, are of non-specialized type, are well nourished, and are highly radiosensitive.
4. Based on the radiation response of normal tissues, they are classified as (1) early-responding tissues and (2) late-responding tissues. Early(acute)-responding tissues consist of fast-proliferating cells and are in turn highly radiosensitive. They may show clinical symptoms within a few weeks of irradiation, for example, bone marrow, spleen, thymus, lymphatic nodes, germinal cells of the ovary and testis, eye lens, lymphocytes, etc. Late-responding tissues have a large number of cells in the resting phase of the cell cycle, so they are lesser radiosensitive. These tissues may take many months or years to develop clinical symptoms. For example, skin, liver, heart, lung, spinal cord, bladder, etc. The comparison of cell survival curves between early- and late-responding tissues is shown in Figure 4.10.
5. From Figure 4.10, it is observed that the survival curve of late-responding tissue has an increased slope than that for the early-responding tissue. This implies that late-responding tissues have greater repair capacity (low α/β ratio) than early-responding tissues (high α/β ratio). In order to avoid late toxicity effects, the dose/fraction is reduced while the number of fractions is increased (hyperfractionation), which will reduce damage to late-responding tissues while maintaining the effectiveness on early-responding tissues or tumors.

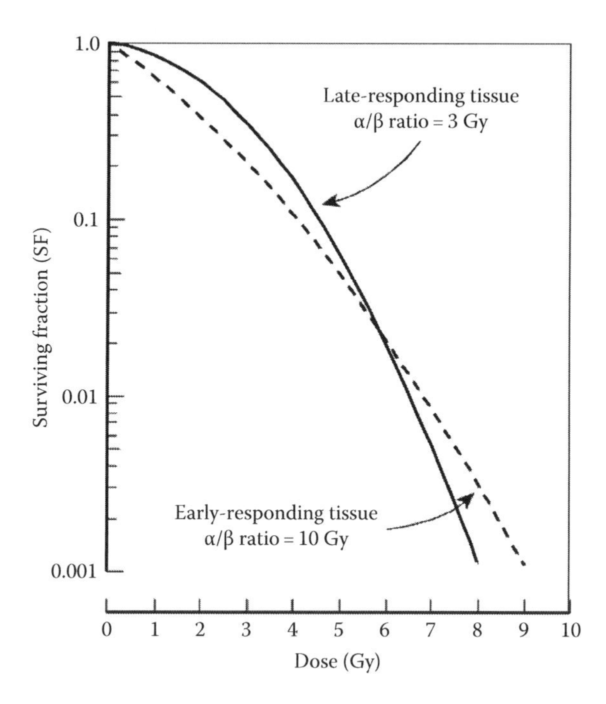

FIGURE 4.10
Survival curve for early- and late-responding tissues. (From Mayles, P. et al., *Handbook of Radiotherapy Physics: Theory and Practice*, CRC Press/Taylor & Francis, Boca Raton, FL, 2007, p. 136, Figure 7.6. With permission.)

6. At higher doses, the survival curve shows higher survival in early-responding tissues than in late-responding tissues because, for early-responding tissues, there is almost complete recovery in a few months so that a second high dose of radiation can be tolerated. For late-responding tissues the extent of residual injury depends on the level of the initial damage and it is tissue dependent [IAEA training course series No. 42]. Hence, the recovery rate of early-responding tissue damage is higher than recovery rate of late-responding tissue damage.

The response to radiation of malignant tissues is generally explained by the number of cells "killed." As per the law of Bergonie and Tribondeau, malignant tissues are more radiosensitive than normal tissues as they have greater reproductive capacity. But, all tumors are not equally radiosensitive as their dividing rate and the fraction of cells present in the resting phase of the cell cycle are varied. For example, lymphomas, leukemia, neuroblastoma, etc. show higher radiosensitivity; squamous cell carcinoma, bladder carcinoma, retinal glioma, etc. show limited sensitivity; and osteosarcoma, soft tissue sarcoma, and melanoma show the least radiosensitivity.

Radiosensitivity of tumors/normal tissues is also influenced by many other factors such as LET, dose rate, oxygen concentration, cell cycle status, repair, etc.

4.3.2 Radiation-Induced Bystander Effect

The radiation-induced bystander effect has been believed to be an important part of ionizing radiation response in the last few years. It is a phenomenon in which unirradiated cells exhibit irradiation effects (such as reduction in cell survival, cytogenetic damage,

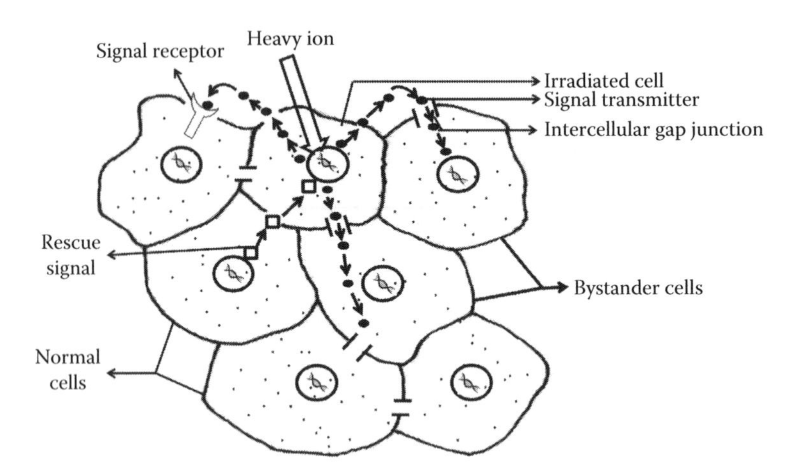

FIGURE 4.11
Schematic representation of the bystander effect.

apoptosis enhancement, chromosome aberrations, micronucleation, transformation, mutation, etc.) as a result of molecular signals received from nearby irradiated cells. Bystander signals may be transmitted either by direct cell-to-cell contact or through soluble factors released into the culture medium (e.g., ROS, cytokines, calcium ions, small RNA, etc.). The schematic representation of the bystander effect is shown in Figure 4.11. A novel research on analyzing the nature of molecular signaling and its related processes is under development. Recently, studies on animals have shown that the bystander effect may also extend to organs and tissues that are far away from the exposed region depending upon the type of tissue. Also, non-irradiated neighboring normal cells may suppress the effect of radiation on irradiated tumor cells by sending rescue signals. In radiation therapy, evidence of bystander events has been observed in the form of abscopal effects (irradiation of one portion of the anatomy affects a portion outside the radiation field). Hence, it is believed that the bystander effect may play a significant role in optimizing radiation therapy and radio diagnostic procedures because it may increase or decrease the clinical outcome depending upon the types of cells receiving those signals. If the extended tumor cells receive signal from the irradiated tumor cells, it may improve the clinical outcome. Nevertheless, if the normal cells receive those signals, it may enhance the side effects such as secondary carcinogenesis, mutations, and genetic instability.

4.3.3 Age

Age is one of the important factors determining the radiation response in terms of varying radiosensitivity. Based on the law of Bergonie and Tribondeau, humans are most radiosensitive before birth and then the radiosensitivity decreases gradually until maturity as shown in Figure 4.12. Although in the case of a child the cells are dividing more frequently than an adult, the same radiosensitivity is observed in old humans, which may be related to low repair capacity at older ages. Hence, radiodiagnosis and radiotherapy for children must be considered more critically than for adult patients. In terms of gender, women are said to be more radiosensitive than men due to hormonal differences. Still, it requires more research with many numbers of patients in different age groups in order to deliver patient-specific treatment.

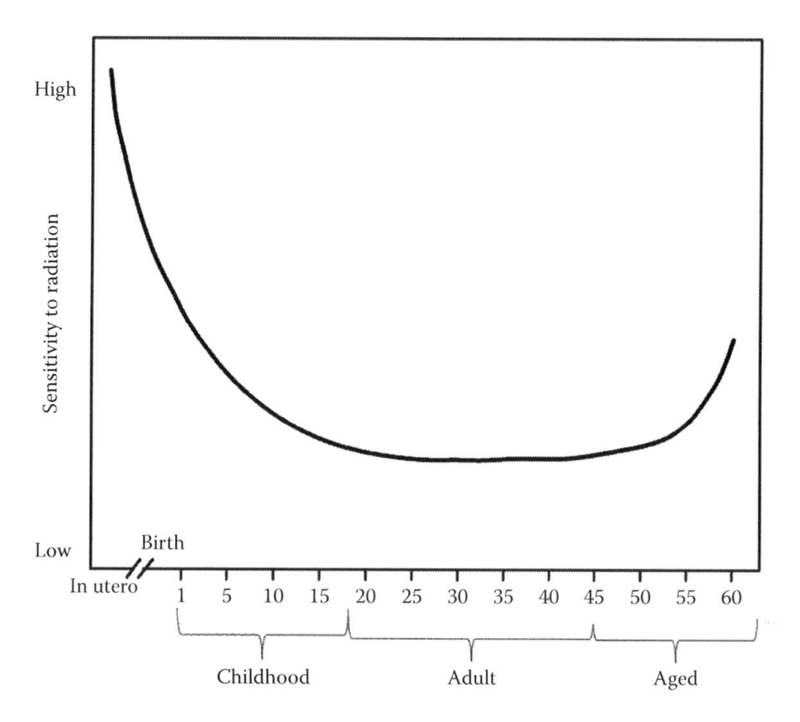

FIGURE 4.12
Age vs radiosensitivity.

4.4 Chemical Factors

4.4.1 Impact of Molecular Oxygen: Oxygen Enhancement Ratio

The radiation-induced biological damage is enhanced when irradiated in the oxygenated (enrich with oxygen) or aerobic/aerated (adequate oxygen) condition than in anoxic (no oxygen) or hypoxic (low oxygen) conditions. As discussed in Chapter 3, when indirectly ionizing radiation (x-rays, gamma rays, and neutrons) interacts with water, hydrogen ($H^\bullet$), organic ($R^\bullet$), and hydroxy ($OH^\bullet$) free radicals are produced. When these free radicals interact with oxygen in the medium, toxic organic/hydrogen peroxide (RO_2/H_2O_2) is formed, which can induce, extend, and fix the cellular damage. Hence, molecular oxygen acts as a potent and nontoxic radiosensitizer. The simple mechanism behind this oxygen effect is given as follows:

$$R^\bullet(\text{free radical}) + O_2 \rightarrow RO_2^\bullet \rightarrow ROOH(\text{toxic hydrogen peroxide})(\text{or})$$
$$CH_2^\bullet(\text{organic free radical}) + O_2 \rightarrow CH_2O_2(\text{organic peroxide, which fixes the damage})$$

Oxygen tension between tissues may vary from 1 to 100 mmHg. Even a very small amount of oxygen can vary the oxygen effect (radiosensitivity) enormously. The dependence of radiosensitivity on oxygen concentration is shown in Figure 4.13. It is observed

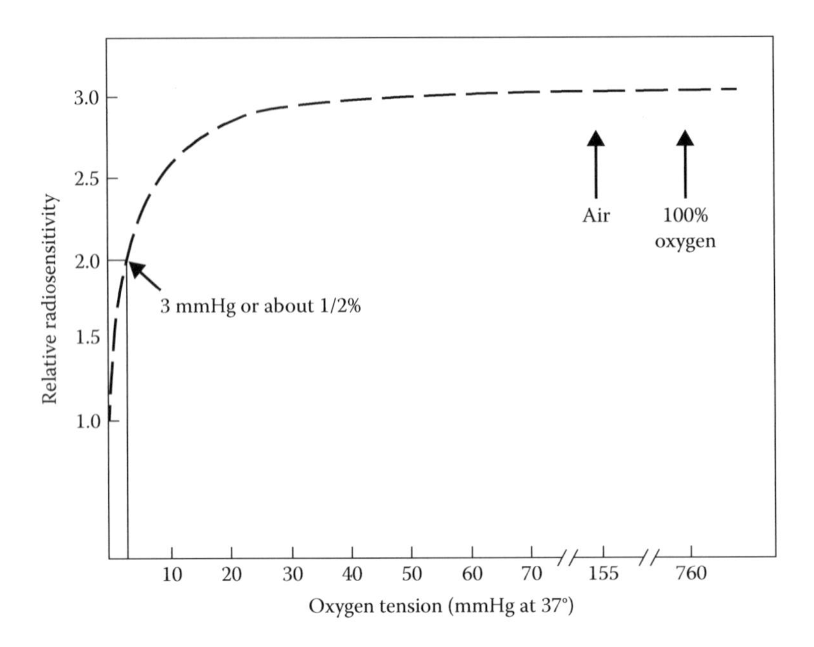

FIGURE 4.13
Dependence of radiosensitivity on oxygen concentration. (From Lehnert, S., *Radio Sensitizers and Radio Chemotherapy in the Treatment of Cancer*, CRC Publisher, 2015, p. 25, Figure 2.2. With permission.)

that most of the change in sensitivity occurs as the oxygen tension increases from 0 to 30 mmHg (5% oxygen). Actually, beyond 2% radiosensitivity at 3 mmHg (0.5%) oxygen concentration, the survival curve is same as that obtained from normal aeration conditions. Further increase in oxygen from air characteristics to 100% oxygen does not affect the radiosensitivity.

The oxygen effect is described numerically by the term "oxygen enhancement ratio" (OER). OER is defined as the ratio of the doses of radiation that produce a given biologic response under anoxic/hypoxic conditions to the dose of radiation that produces the same biologic response under aerobic conditions.

$$\text{OER} = \frac{\text{Dose in hypoxia conditions}}{\text{Dose in oxygenated conditions}}$$

OER depends on LET. The OER is most pronounced for low LET radiation as it produces free radicals (indirect action of radiation) to interact with oxygen, and is less effective with high LET radiation as the damage is beyond repair and can directly hit the target (DNA). For mammal cells, the OER for low LET radiation is between 2 and 3, for medium LET radiation it is between 1.2 and 1.7, and for high LET radiations it is approx. 1. For example, the OER value for x-rays, neutrons, and alpha particles is 2.8, 1.7, and 1.0, respectively. OER also depends upon the time when the oxygen is administered. The oxygen effect is more enhanced when the oxygen is administered simultaneously along with irradiation compared to either before or after irradiation. A plot of the survival curve for cells irradiated to

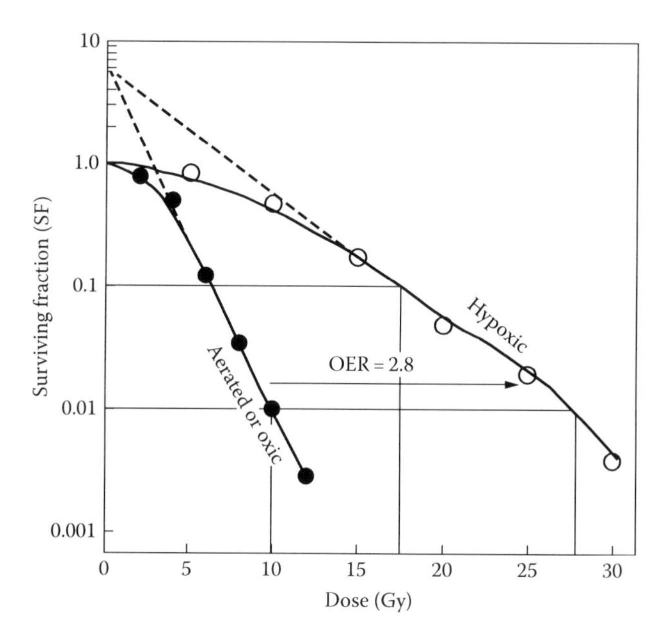

FIGURE 4.14
Survival curve for cells irradiated by x-rays under hypoxic and aerated conditions. (From Mayles, P. et al., *Handbook of Radiotherapy Physics: Theory and Practice*, CRC Press/Taylor & Francis, Boca Raton, FL, 2007, p. 169, Figure 9.4. With permission.)

x-rays under hypoxic and anoxic conditions is shown in Figure 4.14. From this, we can calculate the OER value for x-rays as follows:

We know, $$\text{OER} = \frac{\text{Hypoxic dose}}{\text{Aerated dose}} (\text{or})(\text{Dose in absence/presence of } O_2)$$

From Figure 4.14 for SF = 0.01, $$\text{OER} = \frac{28 \text{ Gy}}{10 \text{ Gy}} = 2.8 \text{ for x-rays}$$

From Figure 4.14 for SF = 0.1, $$\text{OER} = \frac{17.5 \text{ Gy}}{6.2 \text{ Gy}} = 2.8 \text{ for x-rays (the same).}$$

This implies that the OER is independent of dose and biological endpoint (SF) for a particular type of radiation. From Figure 4.14, it is also observed that

1. The shoulder region (Dq) of the survival curve under aerated (fully oxygen) conditions is smaller than for the hypoxic (low oxygen) conditions. This is due to the absence of a repair mechanism in the presence of oxygen. As there is more repair due to lack of oxygen under hypoxic conditions, the shoulder is broader.
2. The slope of the exponential portion of the curve under aerated condition is steeper than for hypoxic conditions, which results in a decreased D_0 dose. This observation implies that oxygen fixes/locks the damage permanently.

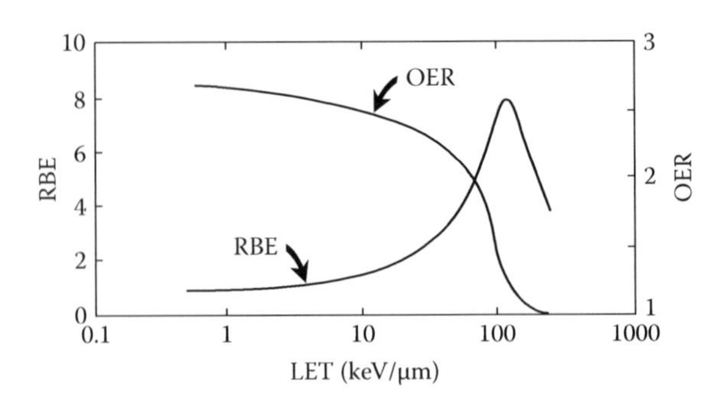

FIGURE 4.15
Relations between LET, RBE, and OER. (With kind permission from Springer Science+Business Media: *Nuclear Medicine Radiation Dosimetry*, 2010, p. 423, McParland, B.J., Figure 10.21.)

The variation in RBE and OER in terms of LET is shown in Figure 4.15. It represents that

1. The RBE and OER curves are mirror images of each other.
2. Rapid increase of RBE and the rapid fall of OER both occur at the same LET of approximately 100 keV/μm.

4.4.2 Chemical Radiosensitizers, Radioprotectors, and Mitigators

The effectiveness of radiotherapy can be improved with the use of chemical substances, such as radiosensitizer, radioprotectors, and mitigators. There are many chemicals available for this purpose, but each has its own applications, mode of interaction, and side effects. These substances modify the indirect action of radiation to perform their roles and hence lesser effective for high LET radiation.

A radiosensitizer is a chemical agent used to sensitize the tumor to irradiation in order to increase cell death. These agents capture free radicals (fixation of free radicals) by their electron affinity molecular structure, and, in turn, prevent the repair of cellular damage like oxygen effect. A few examples of radiosensitizers are

1. Nicotinamide
2. Metronidazole and its analogs such as nimorazole
3. Hypoxic cell cytotoxic agents such as tirapazamine, which sensitize tumor cells by increasing their metabolic activity
4. Radiosensitizing nucleosides, such as fluoropyrimidines (through their metabolites, which leads to cell cycle redistribution, DNA fragmentation, and cell death), thymidine analogs (their incorporation increases the susceptibility of the DNA to single-strand breaks from radiation-produced free radicals), hydroxyurea (through its inhibition of ribonucleotide reductase, a key enzyme for the transformation of ribonucleotides to eoxyribonucleotides), and fludarabine (well-studied DNA damage repair inhibitor)

A radioprotector is a chemical agent used to reduce the radiation-induced damage in normal tissues by reducing genetic effects and intracellular or interstitial oxygen pressure. It consists of a free SH group at one end and a strong basic function, such as amine ($R\text{-}NH_2$) or guanidine (CH_5N_3), at other end. A few of the effective radioprotectors are sulfhydryl compounds like cysteine, mercaptoamines, and amifostine. Cysteine occurs naturally in cells at low levels. If its concentration is increased artificially, then it becomes toxic and causes nausea and vomiting in humans. But, its toxicity can be reduced by covering the SH group of the compound by a phosphate (PO_3) group. Since radioprotectors act as antioxidants, they must be present before or at the time of irradiation. The effectiveness of a radioprotector is expressed by the dose reduction factor (DRF) (ratio of radiation doses producing same biological effects in the presence or absence of radioprotectors) and it is in the order of 1.8. The DRF is also influenced by the type of radiation (low- or high-LET radiation), type of tissue (sensitivity, age, etc.), etc.

A radiation mitigator is also a chemical agent used to prevent acute normal tissue toxicity by interrupting damage pathways, inducing DNA repair mechanisms, and also by triggering normal cells to repopulate. It can be delivered during or shortly after irradiation. Studies shows that the 3,3′-diindolylmethane (DIM) molecule protects rodents from death after potentially lethal damage.

Objective-Type Questions

1. The most sensitive phase of the cell cycle is
   a. G_1 phase
   b. S phase
   c. G_2 phase
   d. M phase
2. The structural organization of spinal cord is
   a. Serial
   b. Parallel
   c. Serial–Parallel
   d. None
3. Molecular oxygen is a
   a. Radiosensitizer
   b. Radioprotector
   c. Mitigator
   d. None
4. One of the following is an early-responding normal tissue.
   a. Heart
   b. Liver
   c. Bone marrow
   d. Kidney
5. To spare late-responding tissues, the following fractionation schedule may be adopted.
   a. Conventional fractionation
   b. Hypofractionation
   c. Hyperfractionation
   d. Split-course treatment

6. The optimum LET to achieve maximum cell killing is
   a. <10 keV/μm
   b. 50 keV/μm
   c. 100 keV/μm
   d. >100 keV/μm
7. Which one of the following cells is the most radiosensitive?
   a. Necrotic cell
   b. Hypoxic cell
   c. Anoxic cell
   d. Aerated cell
8. A human is most radioresistant at this stage.
   a. In utero
   b. Childhood
   c. Adult
   d. Aged
9. The following ionizing radiation shows maximum reduction in cell survival.
   a. X and gamma
   b. Electron
   c. Alpha particle
   d. Thermal neutron
10. Inverse dose rate effect is due to
    a. G_1 block
    b. G_2 block
    c. S phase block
    d. M phase block

5

Biological Effects of Radiation: Deterministic Effects

Objective

To study the various biological effects of radiation on exposed individuals.

5.1 Introduction

According to the International Commission of Radiological Protection (ICRP), the biological effects of radiation can be classified as stochastic and deterministic effects (Figure 5.1). Stochastic effects (random or nondeterministic or probabilistic effects) do not have a threshold dose value and cannot be avoided. They are caused either directly or indirectly by DNA mutations. The severity of a stochastic effect is independent of the absorbed dose. However, as the dose increases, the probability of the effect also increases linearly. Examples are carcinogenesis and mutation. They can be either somatic effects or genetic effects. Somatic effects refer to the biological effects of radiation that are seen on the exposed individual during his/her lifetime itself. In contrast, genetic effects refer to the effects of radiation that are transmitted to the next generation (to the offspring).

Deterministic (nonstochastic) effects are the effects of radiation on exposed individual (somatic effect) that have a threshold dose value below which they do not appear. For example, cataract, blood count changes, erythema (2–5 Gy), infertility, etc., as given in Table 5.1. Above the threshold dose value, the severity of the deterministic effect is increased as the dose is increased. This may be induced by either acute (single high exposure) or chronic (prolonged) exposure.

Based on the time interval between irradiation and its observable effect, deterministic effects may be classified as early effects and late effects. Deterministic effects that can appear in a short time of less than a week after irradiation due to acute exposure are referred to as early effects of radiation. However, late effects are the effects of radiation that take a longer manifestation time in terms of months, years, or sometimes even a few decades after irradiation due to chronic exposure. The severity of deterministic effects also depends upon the region and extent of irradiation, that is, whether it is partial-body irradiation (irradiation of either upper body or lower extremities) or whole-body irradiation. Both of these early- and late-deterministic effects of radiation

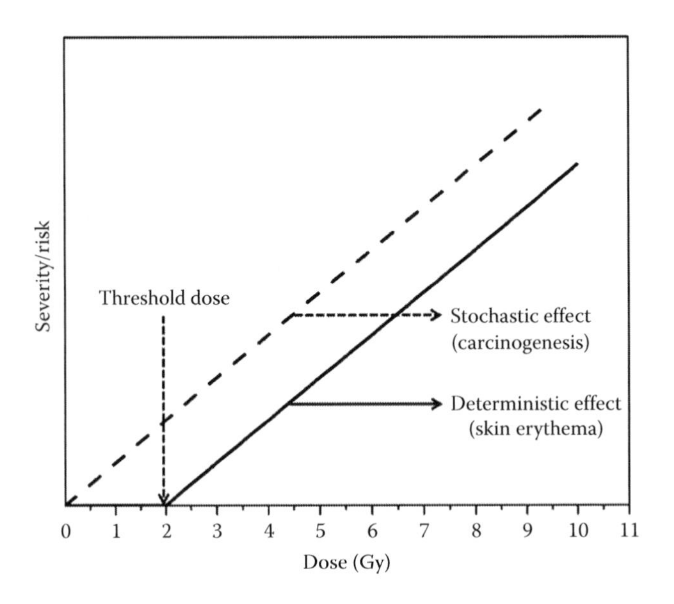

FIGURE 5.1
Radiation dose–response curve.

TABLE 5.1
Threshold Dose Values for Deterministic Effects

	Threshold for Deterministic Effects (1 Sv = 1 Gy for Photons and Electrons)		
Organs	**Effects**	**One Single Absorption (Sv)**	**Prolonged Absorption (Sv-Year)**
Testis	Permanent infertility	3.5–6.0	2
Ovary	Permanent infertility	2.5–6.0	>0.2
Lens of eyes	Milky of lens	0.5–2.0	>0.1
	Cataract	5.0	>0.15
Bone marrow	Blood forming deficiency	0.5	>0.4

Source: With permission from ICRP report No: 103, p. 164, 2007, Table No: A 3.1.

on exposed individuals reported by the ICRP (1984, 1985) and Stewart et al. (2012) and described by Quinn B. Carroll (2011) and Charles A. Kelsey et al. (2014) are discussed in this chapter.

5.2 Early Deterministic Effects of Radiation

Early effects of radiation may appear due to partial- or whole-body irradiation. Early effects of whole-body irradiation include decrease in blood count at a 25-cGy dose, and 50% loss of lymphocytes may appear at a 50-cGy dose. Death of a person may be induced at a 1-Gy threshold dose whereas death of a whole population may occur at a 6–7-Gy threshold dose. In the case of partial-body irradiation, chromosomal aberrations start to manifest even at a 5-cGy dose and extend to erythema (reddening of the skin) at a 2-Gy dose and epilation (hair loss) at a 3-Gy dose, ending up with the failure of the organ/organ system involved, which is discussed in the next section.

5.2.1 Concept of $LD_{50/30}$ and $LD_{50/60}$

Since the various types of human tissues have different radiosensitivities, the effects of radiation among various organs are also varied. Hence, radiosensitivity plays a major role while studying the somatic effects of radiation. Radiosensitivity can be compared among various tissues with various outcomes, namely (1) dose to inhibit mitosis, (2) time taken to regenerate after irradiation, (3) alteration or loss of cellular function, (4) death of cells, (5) time taken to atrophy (removal of cells/tissues), and (6) LD_{50} (mean lethal dose).

LD_{50} is defined as the whole-body mean dose expected to kill (lethal) 50% of a normal population. The related units of LD_{50} are $LD_{50/30}$ and $LD_{50/60}$, where $LD_{50/30}$ is defined as the whole-body mean dose that is expected to be lethal in 50% of the normal population within 30 days, and $LD_{50/60}$ is defined as the whole-body mean dose that is expected to be lethal in 50% of the normal population within 60 days. $LD_{50/60}$ is an important unit to determine the somatic effects of radiation since cell survival beyond 60 days usually results in recovery.

Based on the studies carried out with Hiroshima and Nagasaki nuclear bomb survivors, the $LD_{50/30}$ value was estimated to be in the range of 2.5–4.5 Gy. Later, this dose range was modified into 3.5–4.0 Gy based on the knowledge gathered from the Chernobyl nuclear power plant disaster in 1986. Similarly, $LD_{100/60}$, which is the whole-body mean dose expected to be lethal in 100% of normal population within 60 days, is approximately 7 Gy, even with medical treatment.

5.2.2 Acute Radiation Syndrome

In general, acute radiation syndrome, or radiation sickness, is defined as the radiation syndrome/sickness resulting from the irradiation of a major part of the body (mostly the whole body) by a very high dose of external and highly penetrating radiation (X-rays, gamma rays, neutrons, etc.) in a short period of time (seconds to minutes). It can be induced at a higher dose of above 1 Gy.

The acute radiation syndrome develops due to (1) the release of toxic substances (histamine and mildly toxic histamine-like products) from disintegrated and destructed cells and (2) disturbance of the pituitary–adrenal cortical function. Information regarding the acute radiation syndrome in humans has been collected from (1) the experience of the Japanese at Hiroshima and Nagasaki, (2) the radiation fall-out experience during atomic tests, (3) nuclear reactor accidents, and (4) therapeutic irradiation.

The acute radiation syndrome is different from normal injury as the various stages in acute radiation syndrome after acute exposure and its latent period are different. The latent period is defined as the time period in which hidden histological changes occur from the time of irradiation until the appearance of symptoms. Acute radiation syndrome is a collection of various deterministic effects of radiation that occur in a systematic, sequential manner. These sequential events are called stages of acute radiation syndrome.

5.2.3 Stages of Acute Radiation Syndrome

Early symptoms following irradiation are an effective guide to manage the acute radiation syndrome when physical measurements are not available and the dose received is unknown and nonuniform. These symptoms are associated with four stages of acute radiation syndrome that occur in the following sequential manner.

5.2.3.1 Prodromal Stage (Prodrome)

The prodromal stage starts within minutes or hours after the dose is received and is accompanied by the general symptoms of headaches, vertigo, debility, and abnormal sensations of taste or smell. Then, it extends to the most common gastrointestinal symptoms of nausea, vomiting, and diarrhea, which are commonly known as the NVD syndrome. It may remain for a few minutes to a few days. The severity, time of onset, and duration of this NVD syndrome depend on the dose received. If nausea and vomiting are absent, it may be assumed that the dose was relatively low. If the NVD syndrome occurs immediately and is protracted for a long time, it confirms irradiation of a higher dose.

5.2.3.2 Latent Stage

The NVD syndrome may be accompanied by flu-like symptoms of fever and/or faintness, but they soon subside and the individual often feels a false sense of recovery from illness due to the degeneration and repair of proliferative tissues. Such a period of symptoms represents the latent stage, which also lasts from a few hours to a few days, depending on the dose received. If a person receives a higher dose, the latent period is short and hence this is a good indicator to understand the severity of the syndrome.

5.2.3.3 Manifest Illness Stage

The latent stage is followed by the manifest illness stage in which severe prodromal symptoms come back with the following additional dangerous subsymptoms, which are explained in the next section.

1. Cardiovascular symptoms: Tachycardia (rapid heart rate), arrhythmia (irregular heartbeat), fall of blood pressure, and shortness of breath
2. Hematological symptoms: Leucopenia (decreased white blood cell count), thrombopenia (abnormally smaller number of platelets in the blood), and increased sedimentation rate
3. Psychological symptoms: Increased irritability, insomnia (sleeplessness), and fear

5.2.3.4 Death or Recovery

The manifest illness stage can last from several hours to about two months, during which time the individual may either die or recovers from illness depending upon the dose, volume, and type of tissue irradiated. In general, at higher doses, the predominant syndrome would be the CNS syndrome, which definitely results in death. At lower doses, the gastrointestinal and hematopoietic syndromes would occur. These syndromes may be recovered from by the judicious use of fluid, electrolytes, and their suitable effective treatment under a controlled environment.

5.2.4 General Clinical Subsyndromes of Acute Radiation Syndrome

Based on the predominant, most immediate, cause of death, acute radiation syndrome can be divided into the following three subsyndromes:

1. Hematopoietic or hematological syndrome (bone marrow syndrome)
2. Gastrointestinal syndrome (GI syndrome)
3. Neurovascular syndrome (central nervous system [CNS] syndrome)

5.2.4.1 Hematopoietic Syndrome

The hematopoietic or hematological or bone marrow syndrome occurs when a person is exposed to a 1-Gy to 10-Gy dose. The symptoms during the prodromal stage are anorexia (an eating disorder), nausea, diarrhea, and vomiting that start to occur 1 hour to 2 days after irradiation (onset time) and last for a few minutes to days depending upon the amount of dose.

During the latent stage, stems cells in bone marrow start to die as they are the most sensitive cells as per the law of Bergonie and Tribondeau. However, the exposed person would be normal for 1–6 weeks. Then severe anorexia, fever, and malaise (feeling of discomfort) return and the blood cell counts reduce during the manifest illness stage. The rate of blood cell reduction and its recovery depends upon the dose received.

If a person has received a lower dose, the bone marrow cells start to repopulate the marrow, which leads to the recovery of most of the persons in a few weeks to 2 years after irradiation. If the dose received is 1.2 Gy, death may occur in few persons within a few months; however, the $LD_{50/30}$ value is 3.5–4.0 Gy. The cause of death in the hematopoietic syndrome is infection and hemorrhage (discharge of blood).

5.2.4.2 Gastrointestinal Syndrome

The gastrointestinal or GI syndrome occurs due to the damage to different layers of cells in the intestine when a person is exposed to a high, acute, and whole-body dose of more than 5–6 Gy. The intestine consists of stem cells in its inner crypt layer, intermediate cells, and cells developed into mature cells on the top of the fingerlike villi. The mature gut cells in the villi can get food particles and play their specialized roles, which include absorption of nutrients and the discharge of waste products into the intestine. Since the stem cells are the most radiosensitive cells, they are subject to severe damage. But, the mature cells are radioresistive and hence their functions are not affected due to irradiation. Generally, the differentiation of stem cells into mature gut cells requires about 14 days. Up to 14 days, the mature and intermediate cells of the intestine would be present to perform their function. Thus, the severe effects of intestinal irradiation would appear only after 14 days.

However, the prodromal symptoms of anorexia, severe nausea, diarrhea, vomiting, and cramps (muscle contraction causing severe pain) start to occur within few hours after irradiation and last for about 2 days. During the latent stage, bone marrow stem cells and cells lining the GI track start to die and manifest into anorexia, severe diarrhea, fever, dehydration, and electrolyte imbalance. Finally, it ends up with death within 2 weeks of irradiation since a human can survive only 7 to 10 days without food, and only 4 to 7 days without water. The death by GI syndrome is due to infection, dehydration, and electrolyte imbalance in association with bone marrow syndrome.

5.2.4.3 Neurovascular Syndrome

The neurovascular syndrome, or the central nervous system syndrome, or the CNS syndrome develops when a person is exposed to a very high, acute, and whole-body dose of more than 50 Gy. At this very high dose, a person may experience all the three types of subsyndromes before death. Its prodromal symptoms, such as extreme nervousness, confusion, severe nausea, vomiting, watery diarrhea, loss of consciousness, and burning sensations of the skin occurs within a few minutes after exposure and remains for a few minutes to hours. The irradiated person remains in a stage of partial functionality for just an hour or less and attains coma stage for 5–6 hours after exposure that finally leads to death within 3 days of exposure.

This kind of death mostly occurs due to the destruction of the whole-body circulatory system and increased pressure on the brain due to fluid leakage caused by edema (a condition of abnormally large fluid volume in the circulatory system or in tissues between the body cells), vasculitis (a group of disorders that destroy blood vessels), and meningitis (a serious inflammation of the meninges, the thin, membranous covering of the brain and the spinal cord).

5.3 Late Deterministic Effects of Radiation

In radiation oncology, various radiotherapy procedures deliver a few tens of Gy to a limited region of the body by irradiating both the tumor region and its nearby normal organs and tissues. In these procedures, dose delivery is carefully monitored and recorded, and the effects of radiation on patients are analyzed during and after the end of treatment (follow-up procedure). Based on the information gathered during and after these procedures, the late deterministic effects of radiation on different organs and tissues have been reported by various agencies and researchers as discussed below.

5.3.1 Radiation Effects on Important Organs

Recently, the damage to any organ receiving high dose has been analyzed in terms of two important parameters namely (NTCP) TD 5/5 and (NTCP) TD 50/5. (NTCP) TD 5/5 denotes the tumor dose that has the probability of producing 5% complications in normal tissues within 5 years of the treatment. Similarly, (NTCP) TD 50/5 denotes the tumor dose that has the probability of producing 50% complications in normal tissues within 5 years of the treatment. The (NTCP) TD 5/5 and (NTCP) TD 50/5 values of some important organs and their complications are tabulated in Table 5.2.

5.3.2 Radiation Effects on Organ Systems

The late effects of radiation on the following 11 human organ systems are discussed in addition to the information given in Table 5.2.

5.3.2.1 Circulatory System

The circulatory system includes the heart, veins, and arteries. The heart is radioresistive as it consists of muscle cells that are in the resting phase of the cell cycle. However, protracted

TABLE 5.2

(NTCP) TD 5/5 and (NTCP) TD 50/5 Values of Some Important Organs and Their Complications

Organ/Tissue	(NTCP) TD 5/5 Value	(NTCP) TD 50/5 Value	Complications (Endpoint)
Bladder	65 Gy	80 Gy	Bladder contraction, volume loss, and difficulty in fully emptying or leaking from the bladder
Bone	52 Gy	65 Gy	Necrosis (a type of cell death) and pathologic fractures without any external trauma
Brain	45 Gy	60 Gy	Necrosis and infarction (death of group of tissues due to lack of blood supply)
Esophagus	55 Gy	68 Gy	Necrosis of the lining of the esophagus
Eye	50 Gy	65 Gy	Cataract and blindness
Heart	40 Gy	50 Gy	Pericarditis (inflammation of the sac surrounding the heart), cardiomyopathy (damage of heart muscles), loss of flexibility of heart valves, and coronary artery diseases
Intestine	40 Gy	55 Gy	Obstruction and/or perforation leading to massive infection; nausea, vomiting, and diarrhea
Kidneys	23 Gy	28 Gy	Clinical nephritis (inflammation of the kidney) and kidney failure
Liver	30 Gy	40 Gy	Jaundice and liver failure
Lungs	17.5 Gy	24.5 Gy	Persistent cough and pneumonitis
Rectum, bladder, and prostate	60 Gy	80 Gy	Diarrhea, severe proctitis, necrosis, stenosis, and fistula formation
Spinal cord	47.7 Gy	68.3 Gy	Myelitis, necrosis, and paralysis

Source: Data collected from The International Commission on Radiological Protection, ICRP report, Early and late effects of radiation in normal tissues and organs: threshold doses for tissue reactions and other non-cancer effects of radiation in a radiation protection context, 2011.

delivery of doses above 40–60 Gy induces late changes including endothelial proliferation, thickening of the wall, narrowing of the lumen leading to decreased blood flow, atrophy (gradual loss) of smooth muscles, loss of elasticity of arterial walls, and focal vascular constriction and dilation (Rubin and Casarett, 1968; Law, 1981). As per ICRP 2011, the threshold dose for depression of hematopoiesis is about 0.5 Gy for acute exposure, and 0.4 Gy per year for chronic dose rates.

In addition to the hematopoietic syndrome as discussed in the previous section, late effects are also seen due to the damage to blood vessels, capillaries (extremely small blood vessels), and their endothelial cells. When the blood vessels are damaged, most of the long-term and late effects on tissues and organs occur since blood vessels form the microvasculature of those tissues and organs. Similarly, damage to the capillaries of any organ results in its functional abnormality since these capillaries provide oxygen and nutrients to the cells of the organ.

Anatomically, the interior surface of the blood and lymphatic vessels are lined with a thin layer of endothelial cells. These cells act as a barrier between the blood or lymph inside the vessel and the vessel wall. When the endothelial cells are damaged, white blood cells and fluids are allowed to pass through the vessel.

5.3.2.2 Lymphatic System

When lymphatic vessels and soft tissues are exposed during the conventional treatment procedures (e.g., breast cancer), complex hormonal and oxidative changes take place within the tissue microenvironment. These changes remain for many years and lead to apoptosis, production of free radicals, and change in gene expression that result in increased fibrosis and decreased vascular/lymphatic vessel organization and function. Later, it appears as debilitating lymphedema. It is a chronic debilitating condition that may lead to significant physical and psychological morbidity. If these effects are allowed to progress, they can lead to chronic dermal congestion, fibrosis, decreased limb mobility, pain, and paresthesias (Lawenda, 2008).

5.3.2.3 Respiratory System

The respiratory system includes lungs and the trachea. Even though the respiratory system can tolerate a high level of radiation damage, two types of damage may appear: (1) early pneumonitis within weeks to months after acute exposure and (2) long-term changes related to pulmonary fibrosis within a year. The ICRP, 2011, recommended threshold values for pneumonitis derived from whole-lung radiotherapeutic exposures (usually 5 years of follow-up) of about 6.5 Gy for acute exposures.

5.3.2.4 Digestive System

When the digestive system receives a fractionated dose in the range of 30–70 Gy (Rubin and Casarett, 1968), long-term complications appear in a few months to years after irradiation. A few notable complications are fibrosis, strictures, intestinal perforation, and fistula formation.

5.3.2.5 Endocrine System

Generally, doses in the LD_{50} range show an indication of damage in the endocrine tissue. The hypothalamic, pituitary, and adrenal glands are relatively radioresistant in humans. The thyroid gland is moderately radioresistant in a normal adult as it has a low rate of cell turnover. But, it is the most radiosensitive organ in children. In children, an asymptomatic decrease in thyroid and hypothyroidism accompanied by growth retardation was noted in Marshall Islanders who were exposed to 4 Gy and 7 Gy doses respectively (Larsen et al., 1982). In adults, the threshold dose for myxedema (due to advanced hypothyroidism, or deficiency of thyroxine) is a fractioned dose of 25 Gy.

5.3.2.6 Nervous System

The nervous system is highly radioresistant. However, severe effects can be seen in the brain and spinal cord at higher doses. When the brain receives a fractionated dose of 10 Gy, the symptoms include leukoencephalopathy (a group of diseases affecting the white matter of the brain) and EEG changes, and functional disturbances are recorded within few months. But, at the fractionated dose of 55 Gy, severe symptoms like necrosis, damage to vasculature, and neurological changes are manifested in 1 to 3 years. The spinal cord shows symptoms in 6 months to 2 years. They include myelitis, necrosis of neurons in the white matter, and paresthesia like tingling, weakness, and paralysis due to the damage of fine vasculature. The ICRP, 2011, recommended that the threshold dose for symptomatic spinal cord injury (myelitis) is about 50 Gy delivered in 2-Gy fractions.

5.3.2.7 ***Reproductive System***

The germ cells of the ovary and testes are some of the most radiosensitive cells. In the ovary, the mature oocytes comprise the most radiosensitive germ-cell stage. At doses of about 1.5 Gy, temporary sterility in women occurs. But, higher doses of about 6.0 Gy result in permanent sterility. As the age of a female patient increases, the threshold for permanent sterility is decreased. In male patients, temporary sterility occurs at doses above 2.5 Gy and permanent sterility occurs at doses above 5.0 Gy.

5.3.2.8 ***Skeletal System***

The physiopathology of radiation-induced bone damage is not yet understood clearly. The skeletal system of an adult is relatively radioresistant, so it is difficult to record its late effects. But, absorption of radiation by children's bones may show scoliosis (curvature of the spine), epiphyseal slippage, avascular necrosis, osteoradionecrosis of the mandible, and abnormalities of craniofacial growth after radiation and may finally result in aseptic necrosis and fracture without any external trauma in later life. ICRP, 2011, recommended that the threshold dose for necrosis of femoral heads and fractures of ribs (after 5 years) is around 50 Gy in 2-Gy fractions, and about 55 Gy for skeletal muscles.

5.3.2.9 ***Muscular System***

Joints and muscles of the muscular system may be damaged at a higher level of irradiation although they are highly radioresistant. Due to the damage to tendons and muscles, a joint can become shorter and lose its elasticity. As a result, the joint cannot move through its normal range of motion, and muscular tissue may become tough and painful and lose its strength.

5.3.2.10 ***Urinary System***

At lower doses, the excretory function of the urinary tract is reduced gradually over a period of 6 months to a year. This is due to the degenerative changes in the vasculature and epithelium of nephrons. At higher doses of more than 40 Gy, major symptoms, such as hypertension, albuminurea (the presence of the protein albumin in urea), and renal failure are noticed in a period of 6 months to a year that confirmed the end point of fatal nephritis-like reactions (Law, 1981).

As per the ICRP report of 2011, the kidneys are the most sensitive organs, with the bladder and the ureters being more resistant (deduced from radiotherapeutic experience, with a follow-up time of usually 5 years). The threshold dose for the human kidney is an acute dose of about 7–8 Gy, approaching 20 Gy for doses given as multiple 2-Gy fractions. For late reactions in the bladder and the ureters, the threshold total fractionated (2 Gy fractions) dose is ≤50 Gy.

5.3.2.11 ***Integumentary System (Includes Skin)***

The effects of radiation on the integumentary system include various sequential effects on skin, hair, etc. As per the ICRP report (ICRP, 1991), both deterministic and stochastic effects are observed on the skin. The skin is the largest organ in the body and it is most important one since it is exposed to all teletherapy procedures.

The skin consists of three layers of cells: (1) an outer layer of epithelial cells covered with a layer of protective dead cells, called epidermis; (2) a middle layer of skin consisting of muscle fibers, blood vessels, lymph vessels, sweat glands, and hair follicles, called dermis, and (3) a basal layer of stem cells that will be differentiated into other skin cells in 2–6 weeks' time. The basal layer of cells is responsible for the deterministic effects of radiation since it is more radiosensitive than the upper two layers. However, the epidermis shows the early and acute effects of radiation as it is on the top of the skin and is also made up of rapidly dividing cells. The dermis shows the late effects of radiation.

The sequence of deterministic effects of radiation on the skin is given here:

1. At the threshold dose of about 2 Gy, reddening of the skin (erythema) is seen.
2. Irradiation of the skin at about 3 Gy shows (a) onset of transitory (short time) erythema within hours, and (b) prolonged (long time) erythema followed by epilation (loss of hair in clumps) dry desquamation (skin peeling) in 3–6 weeks.
3. Irradiation of the skin with greater than 20 Gy doses shows moist desquamation (severe skin peeling) and secondary ulceration after 6 weeks.
4. Irradiation of the skin with greater than 30 Gy doses shows dermal necrosis after 10 weeks.
5. At the threshold dose of 30–40 Gy, long-term effects are recorded. They include (a) changes in pigmentation, atrophy of epidermis, sweat glands, and hair follicles, and fibrosis of dermis after 6 months and (b) damage to vasculature including telangiectasia (also called spider veins, which is a benign skin condition) after 1 year.

5.3.3 Induction of Cataract

Cataract is defined as the opacification of the normally transparent eye lens. Anatomically, the eye consists of epithelium cells that can undergo mitotic cell division. These cells are differentiated into lens fibers and attached at the equator of the eye. If there is any spontaneous or radiation-induced abnormality, these cells cannot be differentiated correctly, which results in cataract. Cataract is a deterministic effect of radiation. Initially, it starts as an opacity near the posterior pole of the eye and extends toward the eye lens.

Studies have reported that the conventional fractionated treatment can cause cataract after a latent period of 2 years. For many decades, it was believed that the chronic dose of 2 Gy is the threshold dose that induces cataract. However, recent studies on interventional physicians showed that the threshold for cataract induction may be significantly lower (<0.8 Gy) than the earlier value.

Based on this, the current regulatory limit is set at 150 mGy/year in order to reduce the risk of radiation-induced cataract. Accordingly, ICRP published a report in 2011 saying that the threshold for the absorbed dose for the lens of the eye is considered to be 0.5 Gy. For radiation workers, the ICRP recommends an equivalent dose limit for the lens of the eye of 20 mSv/year.

5.3.4 Radiation Effects on the Developing Embryo or Fetus

The harmful effects of radiation on human embryos were first recorded in 1901–1904. Later, various abnormalities including (1) mental retardation and small head diameter in children born to mothers who received pelvic radiation therapy during pregnancy, (2) mental retardation and microcephaly caused by irradiation at 8–15 weeks of pregnancy in Japanese survivors, and (3) leukemia and childhood cancer in children irradiated in utero with diagnostic X-rays were reported by scientists. Based on these, safety reports paid attention to the hazards associated with pregnant women and considered the fetus to be the most radiosensitive tissue. However, the fetal dose of lower than 100 mGy is taken as a negligible level of concern. Actually, the fetal dose is found to be about 25 % of the dose received on the surface of the abdomen.

The effects of radiation on the developing embryo or fetus due to in utero irradiation are referred to as the "teratogenic effects of radiation." It represents the hazardous effects induced by irradiation of a woman after conception. Although the dose–response relationship of the teratogenic effects of radiation is not well understood, it is believed that it has a nonlinear relationship with the threshold. The severity of the teratogenic effects depends on the gestational age of the fetus, dose, and the dose rate.

Intrauterine development in humans can be divided into three phases: (1) period of blastogenesis, (2) period of major organogenesis, and (3) period of fetogenesis. The developmental processes of embryo through these periods and their effects during these periods are discussed here briefly. At this junction, the starting and ending days of these periods should not be taken as fixed cut-off points since the effects of radiation on the embryo between these periods are overlapped.

5.3.4.1 Period of Blastogenesis

When the zygote moves through the fallopian tube toward the uterus, it undergoes cleavage and first division takes place in 24 to 30 hours. Then, it starts to divide rapidly to form a multicellular blastocyte. It floats in the uterine cavity for about 5 days and becomes fertilized. Then, the fertilized blastocyte starts to embed in the uterine wall for implantation. The time between the fertilization and the implantation of blastocyte in the uterine wall is referred to as the period of blastogenesis or the preimplantation period. This period ranges from 0 to 2 weeks postconception where the 0th day is considered to be the day of successful mating. At this stage, cell proliferation takes place effectively.

If the blastocyte is exposed to radiation during this period, it may either abort spontaneously or survive without any observable effects of radiation. This can be represented as an "all or none" phenomenon. Studies estimated that the dose of 10 cGy increases the risk of spontaneous abortion from a normal rate of 25 % to just 25.1 %.

5.3.4.2 Period of Major Organogenesis

After implantation, toward the process of embryonic development, organ formation takes place over a period from 2 to 7 weeks post-conception. Hence, this period is referred to as the period of organogenesis. During this period, intensive cell proliferation, differentiation, and migration takes place, which results in the formation of precise morphological structures. By the end of this period, all the 11 organ systems are established and its external body features are formed in a recognizable manner.

Irradiation of the fetus during this period results in various congenital abnormalities and growth retardation. Specifically, irradiation during the earlier organogenesis period leads to various skeletal defects such as a stunted limb, a missing or extra finger, extended coccygeal segment, or other structural malformation. Irradiation of the fetus during its later organogenesis period results in various neurological and motor deformities, such as improper eye formation, improper cerebrum formation, and failure of the formation of the cranial vault over the brain.

5.3.4.3 Period of Fetogenesis

During the period of fetogenesis from 8 to 38 weeks (266–270 days of total gestation time) the fetus grows in size and all the organs are differentiated further. By the end of this period, an adult morphological feature is fully developed. But, a few organs like brain continue its developmental process even after birth.

Upon irradiation at a lower dose of 5–50 cGy during the 8–16 weeks of gestation in its fetogenesis period, the fetus is most susceptible to growth retardation, such as a small head size, mental retardation up to 20 %, and reduction in IQ (Intelligence Quotient), but the possibility of abortion is reduced. Because the brain is rapidly developing during this period the reduction in mental acuity (sharpness of vision, and the visual ability to resolve fine details measurable by a Snellen chart) has been noted only in doses above 100 mGy. Studies estimated that the reduction in mental acuity is 21–29 IQ points per Gy out of the normal IQ score of 107 points. As the fetal dose is increased, the severity of all the low-dose effects is increased in addition to the possible incidence of miscarriage and other malfunctions.

A fetal dose of more than 50 cGy during the 16–25 weeks of gestation increases the incidence of miscarriage, growth retardation, reduction in IQ, mental retardation, and other major malfunctions. Exposure of the fetus to high dose irradiation during the 26–38 weeks of gestation may also result in neonatal death (De Santis et al., 2007).

In addition to all deterministic effects, low-dose irradiation of the fetus during the late fetogenesis period can cause an increased risk of childhood malignancies, principally leukemia. Various studies confirmed that a dose of around 1 cGy can increase the risk and the excess absolute risk is approximately 6 % per day.

5.3.5 Shortening of Life Span

Studies on mammals confirmed that a substantial amount of either whole-body or partial-body irradiation shortens the life span. But, there are no sufficient data to assess the effect of radiation on the human life span. Studies on U.S. radiologists also showed that the occupational exposure may have a slightly increased mortality rate, but the dose–response relationship could not be measured since the cumulative dose was not known.

Studies on British radiologists and doctors showed no evidence for increase in mortality. Above all, lifetime shortening was not recorded for Japanese atomic bomb survivors, radium watch dial painters, x-rayed patients, and other exposed human populations.

The theory of radiation hormesis suggests that low levels of radiation, which are lower than 10 cGy, are good for us because such low doses may provide a protective effect by stimulating molecular repair and immunologic response mechanisms. However, this theory is not confirmed.

Recent studies estimated that occupational workers may lose about 12 days of life and approximately 10 days for every cGy. But, the estimated average lifetime shortening for other risky conditions are occupational accidents 74 days, general accidents 435 days, one pack of cigarettes in a day 1600 days, being unmarried 2000 days, heart disease 2100 days, and being male rather than female 2800 days.

Objective-Type Questions

1. The fetus is most sensitive to radiation-induced mental retardation during the _____ week period of pregnancy.
   a. 0–7
   b. 8–16
   c. 17–24
   d. 25–36
2. During the most sensitive period for radiation damage, the decrease in mental acuity is estimated to be about _____ IQ points per 1 cGy.
   a. 0.3
   b. 3
   c. 30
   d. 300
3. Erythema occurs at doses of about
   a. 0.02 Gy
   b. 0.2 Gy
   c. 2 Gy
   d. 25% of the X-ray tube output.
4. The main cause of death from the hematopoietic syndrome is
   a. hypotension arising from microvascular destruction.
   b. hemolytic anemia.
   c. infection and hemorrhage resulting from loss of white cells and platelets.
   d. loss of erythrocytes resulting in organ ischemia.
5. The mean lethal dose for humans is about _____ Gy.
   a. 0.5
   b. 5
   c. 50
   d. 500
6. Death is expected within 2 weeks following at whole-body doses of ___ Gy or greater.
   a. 0.1
   b. 1
   c. 10
   d. 100
7. About _____ Gy will produce permanent sterility in men.
   a. 0.2
   b. 2
   c. 0.5
   d. 5

8. The major complication observed in the spinal cord is
   a. Myelitis
   b. Necrosis
   c. Paralysis
   d. All
9. Cataract is _______
   a. Deterministic effect
   b. Stochastic effect
   c. Both
   d. None
10. An occupational worker may lose about _____days of life and approximately ____ days for every cGy
    a. 12 weeks, 10 weeks
    b. 74 weeks, 12 weeks
    c. 12 days, 10 days
    d. 74 days, 12 days

6

Biological Effects of Radiation: Stochastic Effects—Carcinogenesis

Objective

The objective of this chapter is to review the stochastic effects of radiation on the induction of carcinogenesis as a function of dose, as this is the major concern with regard to the safety of radiation workers as well as in the management of patients undergoing diagnostic procedures and radiation therapy.

6.1 Experience on Radiation Carcinogenesis

Carcinogenesis is one of the stochastic effects of radiation, and hence there is no threshold dose. This means that even for a very small dose there is some effect. However, in the case of cancer induction due to radiation exposure, it is the probability effect. This means that at low doses the probability of induction of cancer is small and the probability of induction of cancer increases as a function of dose.

Carcinogenic effects of radiation have been known since the discovery of radioactivity by Becquerel. Both Dr. Marie Curie and her daughter Irene Curie who performed research with radioactive isotopes died of cancer and it was thought that radiation had induced cancer in both of them.

The knowledge of radiation-induced cancer was better understood from the following historical evidences:

- Hiroshima–Nagasaki bomb survivors who have both leukemia and solid tumors
- Patients irradiated for ankylosing spondylitis who have leukemia
- Thyroid cancer observed in children irradiated for enlarged thymus and epilated for tinea capitis
- Breast cancer observed in patients treated with X-rays for postpartum mastitis and patients who took fluoroscopy repeatedly during tuberculosis management
- Lung cancer observed in uranium miners
- Bone cancer observed in radium dial painters who ingested radium and also in patients who were treated with radium for tuberculosis or ankylosing spondylitis

- Skin cancer observed in early X-ray workers who worked around accelerators before implementing the radiation safety standards
- Liver tumor detected in patients who used Thorotrast contrast material, which contains radioactive thorium
- Second cancers reported in patients who were treated for prostate cancer, carcinoma of the cervix, Hodgkin's lymphoma, breast cancer, carcinoma of the testes, and various childhood cancers

6.2 Radiation Epidemiology

Epidemiology refers to the study of distribution of diseases and their causative factors in the human population. Epidemiological studies become necessary when an agent causes the effect only in a small fraction of the individuals exposed to it. This method of assessing the effect is more suitable in radiation epidemiology for the induction of cancer as it is stochastic in nature.

The aim of **radiation epidemiology** is to identify, understand, and quantify the risk of cancer in a population when it is exposed to medical, occupational, or environmental radiation. Medical scientists and epidemiologists are interested in investigating radiation-induced cancer by various epidemiological studies, such as (1) cross-sectional study, (2) ecological study, (3) case control study, and (4) follow-up study. The cross-sectional study (also called the prevalence study) is a simple study performed with the objective of determining the occurrence of cancer at a given point in time. It may be either descriptive or analytic. It can determine both cancer and its corresponding exposure simultaneously. The ecological study is performed on groups (schools, factories, countries, etc.) to measure the exposure as an overall group index. In the case control study, an abnormal population is compared with a normal population to characterize the cancer. The follow-up study (also called cohort study) is performed by comparing the cancer in an exposed group with the cancer in an unexposed group after a follow-up period.

There are many confounding factors that may interfere with these studies, which need to be adjusted. The chief confounding factors are exposure to other agents that are surrogate to the agent under study such as smoking, alcohol intake, infection, medication, and genetic disposition. A good example of a confounding factor is cigarette smoking in lung cancer studies in uranium miners. Another example is the study of prenatal effects of radiation where maternal malnutrition, maternal age, fetal hypoxia, infections, and alcoholism are the major confounding factors. The statistical method used should also be appropriate. For example, if a disease is occurring at low frequency, one can observe a statistical increase; however, if the disease is occurring at a higher frequency in the population, it will be difficult to see a small increase in the disease statistically. Thus, the analysis of the results obtained in epidemiological studies should be conducted with caution.

The committees that assess the health effects of radiation are the International Commission on Radiological Protection (ICRP), Medical Countermeasures against Radiological and Nuclear Threats (NIAID), Health Protection Agency (UK), International Atomic Energy Agency (IAEA), International Agency for Research on Cancer, National Council on Radiation Protection and Measurements (NCRP), United Nations Scientific Committee on the Effects of Atomic Radiation (UNSCEAR), World Health Organization (WHO), and Health Physics Society (USA).

6.3 Linear Nonthreshold (LNT) Hypothesis

The LNT hypothesis states that "above the prevalent background dose, an increment in dose results in a proportional increment in the probability of incurring cancer of ≈0.005% per mSv of dose."

According to the LNT hypothesis, if the estimated risk from 100 mGy to 10,000 people is 50 excess cancers, then

- The risk from 10 mGy would be only 5 excess cancers
- The risk to 100,000 people would be 50 excess cancers
- As would that for 1 mGy to 1,000,000 people, or for 0.1 mGy to 10 million people

If an agent has a suspected risk of causing harm to the public, the burden of proof that it is not harmful falls on those overseeing the agent. Therefore, use of the LNT model is expected to "safely" address risk when observable high-dose radiation effects are extrapolated downward to low doses given at low dose rates. However, the LNT hypothesis has the following issues:

- The adaptive response can substantially modify the effects of radiation.
- The bystander (off target) effect has been amply demonstrated.
- Repair of radiation-induced mutational events is efficient.
- Apoptosis reduces radiation-induced genetic damage.
- Hormesis (beneficial effects) has been demonstrated in epidemiological studies conducted in high background areas like Kerala and in laboratory animal studies.

6.4 Dose and Dose Rate Effectiveness Factor (DDREF)

Since the biological system can repair a significant part of radiation damage for low-intensity (chronic) exposures, it is assumed that the risk should be half for low doses and when radiation is delivered at low dose rates. This is referred to as the dose and dose rate effectiveness factor (DDREF).

6.5 Cancer Risk Estimation

Low-level radiation effects are difficult to perceive for a number of reasons. None of the deterministic effects can occur following chronic radiation exposure for decades. The late stochastic effects, namely carcinogenesis, take years or decades to manifest after exposure. There are problems in detecting the effect mainly because it is probabilistic in nature as it occurs only in a very small fraction of the exposed population and also it is not distinguishable from spontaneously occurring cancer. Moreover, radiation is not a strong carcinogen compared to certain carcinogenic chemicals and environmental agents. In addition, the natural background frequency of cancer is so high (10%–20% mortalities) in human beings that a small increase due to low-level radiation (occupational radiation) cannot be detected. However, one can infer that low-level radiation may be associated with the induction of cancer as it is observed that in a population exposed to higher levels of radiation, there is an increase in the incidence of cancer. Though this infers that there is an association of induction of cancer with low levels of radiation, the data obtained from around the world for a population exposed occupationally in prescribed limits do not show a statistically significant increase in the incidence of cancer.

Most data on carcinogenesis in humans were gathered from a small number of persons who were exposed to larger doses. These data would not be useful to estimate the risk at lower doses and for other criteria. Hence, it is necessary to adopt a model that will include the following factors:

- To extrapolate the high-dose data to the low doses of public health concern
- To estimate the risk for full lifetime and for future using the data collected for a limited period
- To estimate the risk for all population using the data available for Hiroshima–Nagasaki bomb survivors

To do this, two types of models are available: (1) an absolute-risk model and (2) a relative-risk model. The absolute-risk model assumes that radiation induces a discrete dose-related "crop" of cancer over and above the spontaneous level. This model is well fitted to analyze leukemia. The relative-risk model assumes that the effect of radiation is to increase the spontaneous occurrence at all ages following irradiation by a given factor. Since the spontaneous occurrence of cancer is significantly high in old age, this model can predict a large number of excess cancers late in life after irradiation.

Recently, the time-dependent relative-risk model was chosen by the BEIR V committee to assess cancer risk in the Japanese atom bomb survivors. This model considered various factors such as dose, square of the dose, age during exposure, gender, and time since exposure to assess the excess incidence of cancer.

However, various uncertainties occurred while estimating the cancer risk (Evidence Report: Risk of Radiation Carcinogenesis, 2016):

- Uncertainties associated with record keeping, statistics, number of data, misreporting of cancer deaths, and so on
- Uncertainties in the shape of the dose–response curve at low doses (i.e., linear, linear quadratic) and the possibility of dose thresholds
- Uncertainties associated with extrapolation of experimental data from animals to humans
- Uncertainties associated with individual sensitivity, age, gender, genome, dietary, life style, immunity, and so on
- Uncertainties in their models and dosimetry methods

Based on the reports of the UNSCEAR and BEIR V committees, the ICRP suggested that the risk for excess cancer mortality in a working population is 8×10^{-2} per Sv at high dose and high dose rate, and 4×10^{-2} per Sv at low dose and low dose rate.

6.6 Types of Cancer Caused by Radiation Exposure

Irradiation to low-LET radiation can cause lung, colon, rectal, breast, stomach, liver, brain, ovarian, esophageal, and bladder cancers, and several types of leukemia, including acute lymphocytic leukemia, acute myeloid leukemia, and chronic myeloid leukemia (Evidence Report: Risk of Radiation Carcinogenesis, 2016).

Different types of cancer take different time intervals to show their abnormalities. This can be represented in terms of the **latency period**. It is referred to as the time interval between irradiation and the appearance of the abnormality. Generally, leukemia has the shortest latency period of 1–5 years. Thyroid cancer has a latency period of 10–20 years, and other solid tumors have a latency period of 20 years or more.

6.7 Second Cancers in Radiotherapy Patients

Second malignancies arise for three main reasons: (1) continued lifestyle, (2) genetic susceptibility, and (3) induced by radiotherapy.

Most radiation-induced cancers occur in organs and tissues in the high-dose volume but some also appear in the low-dose (<2 Gy) volume. There are pronounced differences in the types of radiation-induced second cancers between children, young adults, and elderly patients treated with radiotherapy. Peak incidence occurs 5–20 years after exposure and depends on the patient's age at the time of treatment. Moreover, the types of second cancers after radiotherapy are different from those induced by low-dose total body irradiation (e.g., Japanese atom bomb survivors).

The risk of radiotherapy-induced cancer is well below 1% after radical radiotherapy of most adult cancers, such as cervix and prostate cancer. The risk of dying from uncontrolled local recurrences within a few years after radiotherapy is much higher than the risk of developing a second cancer 10 or 20 years later.

In pediatric patients, several large studies have demonstrated that both radiotherapy and chemotherapy, and in particular the combination of both, cause a significant risk of developing a second malignancy. Leukemia predominates in the first 10 years whereas various solid cancers develop later in life. The latter risk increases steadily with increasing survival; therefore, a major effort is needed to identify those factors that determine the size of risk. For radiotherapy, it is, in the first instance, the dose and the dose distribution, as well as the volume irradiated. Many studies have been published in recent years that have determined the radiation doses within and outside the target volume as part of the treatment-planning optimization process.

The findings of the epidemiological studies in patients treated for cervix and prostate cancers suggest that two different mechanisms, leading to radiation-induced second cancers, may exist that show very different relationships with radiation dose. One mechanism is related to chronic and consequential radiation damage after high radiation doses and the other is associated with atrophy due to chronic inflammation. Different mechanisms or radiation carcinogenesis at low-radiation doses, for example, in organs outside the target volume, become critical. The most critical organs in the low-dose volume with regard to radiation-induced second cancers are the lung and the stomach.

In contrast to the organ-weighting factors proposed in the ICRP model (ICRP, 2007), the organs shown in the postradiotherapy studies to be at highest risk are very different: In none of the studies has the colon been found to be critical, and in many studies the breast was by far the most critical organ. Therefore, the risk of radiation-induced cancers from radiotherapy cannot be estimated using the effective dose method proposed by the ICRP for radiation protection purposes. Due to different biological mechanisms after radiotherapy, depending on dose distribution and age of the irradiated patient, the dose–risk relationship is unlikely to follow a simple mathematical model.

The risk of radiation-induced cancers may increase by about 0.5% (Hall and Wuu, 2003), when new sophisticated techniques, such as intensity modulated radiation therapy (IMRT), are applied in radiation treatment. The total extra cancer risk posed by IMRT is the sum of radiation due to the extra volume of normal tissue exposed and the total body dose due to extra leakage resulting from a doubling of the number of monitor units.

6.8 Cancer Risks from Diagnostic Radiology

A few years after X-rays were first used for radiologic imaging, physicians and other medical radiation workers developed skin carcinomas, leukemia, dermatitis, cataracts, and other adverse health effects. Despite early recommendations to decrease stray radiation to the patient and restrict the X-ray beam, years passed before these recommendations were implemented and radiation protection committees were established. With the development and evolution of measures of radiation dose, film badge monitoring, and personal (e.g., lead aprons) and general (e.g., lead shields) radiation protection equipment, occupational doses declined dramatically and the excesses of leukemia, skin cancer, and female breast cancer in medical radiation workers employed before 1950 were no longer apparent in subsequent medical radiation workers.

From 1956 to the present, epidemiological studies have also linked diagnostic X-rays with cancer increases in patients, including modest excesses of pediatric leukemia in the offspring of mothers undergoing diagnostic X-rays during pregnancy, and increased breast cancer risks in women with tuberculosis who were monitored using fluoroscopy and in women with scoliosis who were evaluated with repeated X-rays. During the past 30 years, newer imaging modalities (such as computed tomography [CT], myocardial perfusion scans, positron emission tomography [PET], and other radiologic procedures) dramatically increased. These procedures have provided immense clinical benefit but also higher ionizing radiation exposures to patients. Medical radiation now comprises almost 50% of the per capita radiation dose, compared with 15% in the early 1980s. Although the individual risk of developing radiation-related cancer from any single medical imaging procedure is extremely small, the substantial increase in the per capita effective dose between 1980 and 2016, as well as reports of a substantial fraction of patients undergoing repeated higher-dose examinations, motivates the development of diagnostic reference levels (DRLs) and quality assurance (QA) procedures to limit this dose in order to reduce the probability of the occurrence of cancer.

The *British Journal of Radiology* publication (2008) on cancer risks in diagnostic radiology says, "Collective dose from radiological procedures includes contributions from the many low-dose procedures, such as routine chest X-rays and mammograms, which involve doses far below those for which we have direct evidence of cancer induction," i.e. it assumes validity of the LNT hypothesis down to the lowest doses. However, most of the collective dose is actually from high-dose procedures (e.g., CT, interventional radiology and barium enema).

Risk estimates based on effective dose are highly generic and include, for example, hereditary effects that are unlikely to be significant at doses relevant to diagnostic radiology. In addition, the weighting factors used in the calculation of effective dose do not take into account the strong variations of radiosensitivity with age and gender.

Recent epidemiological results suggest an increase of cancer risk after receiving computed tomography (CT) scans in childhood or adolescence. A study reported in *British Journal of Cancer* (2015) suggests that the indication for examinations, whether suspected cancer or predisposing factors (PF) management, should be considered to avoid overestimation of the cancer risks associated with CT scans.

6.9 Attributable Lifetime Risk

To express the health impact of whole-body exposures to radiation, the lifetime risk of total cancer, without distinction of site, is usually of primary concern. Estimates of risk for both mortality and incidence are of interest, the former because it is the most serious consequence of exposure to radiation and the latter because it reflects public health impact more fully. The time or age of cancer occurrence is also of interest, and for this reason, estimates of cancer mortality risks are sometimes accompanied by estimates of the years of life lost or years of life lost per death. Because leukemia exhibits markedly different patterns of risk with time since exposure and other variables, and also because the excess relative risk for leukemia is clearly greater than that for solid cancers, all recent risk assessments have provided separate models and estimates for leukemia.

Lifetime attributable risk (LAR) is the sum of the exposure age plus latency to LAR upper limit of the yearly target population (wEAR) multiplied by the probability of surviving to attained age (a) divided by survival to exposure age (e):

$$\mathrm{LAR} = \sum \mathrm{wEAR}_{\mathrm{target}} \times \frac{\mathrm{Survival}\,(\mathrm{a})}{\mathrm{Survival}\,(\mathrm{e})}$$

where wEAR is the weighted expected attributable risk.

Data in the atom bomb population were not collected until year 5 after exposure; the leukemia rate from 2 to 5 years after exposure is assumed to be equal to the rate 5 years after collection. LAR upper limit is defined as 100.

Population attributable risk (PAR) is the reduction in incidence that would be observed if the population was entirely unexposed compared with its current (actual) exposure pattern. Risk coefficients representing the lifetime radiation-induced cancer mortality (or incidence) attributable to an exposure to ionizing radiation have been published by major international scientific committees. The calculations involve observations in an exposed population and choices of a standard population (for risk transportation), of suitable numerical models, and of computational techniques.

A more realistic perception of lifetime risk could be gained by the use of coefficients scaled to the lifetime spontaneous cancer rates in the standard population. The resulting quantity of lifetime fractional risk is advantageous also because it depends much less on the choice of the reference population than the LAR.

Radiation exposure from the use of CT in the evaluation and management of severe traumatic brain injury causes negligible increases in LAR of cancer and cancer-related mortality.

Estimates of LAR of any cancer for an exposure at age ≤ 40 years were lower in males than in females for any given quantile. At age ≥ 50 years, LAR was similar for both sexes only at the lowest exposure doses, whereas at higher dosage, it was, in general, higher for women. At the median age of this case series (62 years) and for a radiation exposure ranging from 1.33 to 3.81 mSv, LAR was 1 in 4329 (or 23.1 per 105 persons exposed) and 1 in 4629 (or 21.6 per 105 persons exposed) in men and women, respectively. For an exposure ranging from 10.34 to 18.97 mSv at the same median age, the LAR of cancer incidence was 1 in 1336 (or 74.8 per 105 persons) in men and double, that is, 1 in 614, (162.8 per 105 persons) in women.

Objective-Type Questions

1. Carcinogenesis is a
   a. Stochastic effect
   b. Somatic effect
   c. Genetic effect
   d. All of the above
2. Latent period for Leukemia is
   a. >20 years
   b. 10–20 years
   c. ~10 years
   d. 1–5 years
3. ____cancer was observed in uranium miners
   a. Lung
   b. Skin
   c. Liver
   d. All of the above
4. According to the LNT hypothesis, if the estimated risk from 100 mGy to 10,000 people is 50 excess cancers, then the risk to 100,000 people would be
   a. 5 excess cancers
   b. 50 excess cancers
   c. 500 excess cancers
   d. None

5. The ICRP-suggested risk for excess cancer mortality in a working population at low dose and low dose rate is
   a. 8×10^{-2} per Sv
   b. 8×10^{-4} per Sv
   c. 4×10^{-2} per Sv
   d. 4×10^{-4} per Sv
6. The risk of radiotherapy-induced cancer is
   a. <1%
   b. 1%–5%
   c. 5%–10%
   d. >10%
7. Medical radiation now comprises ___ of the per capita radiation dose
   a. ~10%
   b. ~30%
   c. ~40%
   d. ~50%
8. The estimated lifetime attributable risk for women is
   a. 1 in 614
   b. 1 in 1336
   c. 1 in 500
   d. 1 in 1000
9. Which of the following examples is stochastic effect
   a. Nausea and vomiting
   b. Expiation and fatigue
   c. Diarrhea and leukopenia
   d. Cancer and genetic effects
10. The following cancer occurs due to small exposure received over a long period
    a. Leukemia
    b. Cervical cancer
    c. Lung cancer
    d. Colon cancer

7

Biological Effects of Radiation: Stochastic Effects—Genetic Effects

Objective

The objective of this chapter is to study the various biological effects of radiation expressed on the progeny of an exposed individual.

7.1 Genetic Effects of Radiation: Introduction

When the genes in the reproductive cells of a person are mutated by ionizing radiation, the damage passes on to the individual's generation/progeny/offspring. Hence, it is referred to as genetic effects of radiation. These effects are also induced by chemicals to a greater extent than by ionizing radiation. This is one of the stochastic effects of radiation and therefore it has no threshold dose value. As the dose received by a person is increased, the probability of occurrence of these effects also increases.

The genetic effects of X-rays were initially identified by H.J. Muller in 1927 when he performed experiments with fruit flies (*Drosophila*). Based on this, serious efforts were made to analyze the genetic effects that developed in the (1) nuclear industry, (2) radioactive fallout during nuclear weapon test, and (3) diagnostic and medical uses of ionizing radiation, after the Second World War. Later, the National Academy of Sciences–National Research Council (NAS-NRC) established various committees, namely the Biological Effects of Atomic Radiation (BEAR) committee, the Biological Effects of Ionizing Radiation (BEIR) committees I to VII, and the United Nations Scientific Committee on the Effects of Atomic Radiation (UNSCEAR). This chapter discusses the basics behind the genetic effects of radiation consolidated from the report of UNSCEAR (1958, 2001), BEIR V (1990), BEIR VII (2006), *Radiobiology for the Radiologists* (2011), and Asimov and Dobzhansky (1966).

7.1.1 Mutations

It is well known that each and every individual has unique features and characteristics that are neither seen in the past nor will be seen in the future. This is due to the variation in chromosome combination during meiosis with or without crossover. It is reported that a single married couple could produce children with any of 100 trillion possible chromosome combinations. They may also produce specialties in an individual's character since the genes are dependent on each other. If the chromosomes are subjected to more serious structural or chemical changes due to various factors (chemical, radiation, environmental factors, etc.), an individual with entirely new characteristics that were not seen earlier is produced. These changes are referred to as mutations: for example, a child born with color-blindness, which was not present in the family history. Nevertheless, a child with blue eyes born to parents both having brown eyes is not considered to be a mutation because the grandparents or ancestors might have had blue eyes.

Mutations may be classified as chromosome mutations, gene mutations, somatic mutations, and hereditary/genetic mutations. Chromosome mutations are referred to as mutations associated with changes in chromosome structure, which can be visible under the microscope. Examples are (1) polyploidy (many chromosomes in a cell) due to chromosome replication without cellular division, which will lead to quick death in mammals and (2) Down syndrome due to a break in chromosomes and their abnormal reunion where cells consist of 47 chromosomes instead of 46 due to the addition of one chromosome with the 21st chromosome pair. Gene mutation is referred to as the aberration observed in the chemical structure of the genes that make up the chromosome. This includes a series of changes in DNA such as deletion, insertion, rearrangement, base alternations, and breakage in the sugar–phosphate backbone. Somatic mutations are when either chromosome or gene mutations occur on the somatic cells, which will lead to the production of abnormal new cells.

Mutations observed on reproductive/germ cells are referred to as genetic or hereditary mutations or genetic diseases. The reproductive cells in males are spermatozoa. They form when the spermatogonial stem cells in the seminiferous tubules of the testes undergo several stages of development starting from primary spermatocyte, secondary spermatocyte, spermatids, ending with spermatozoa (mature sperm cells). It takes approximately 6 weeks in mice and 10 weeks in humans, and their production, division, and development are continuous from puberty to death. In females, oocytes are the reproductive cells. About millions of oocytes are formed from oogonial stem cells (through mitosis) 3 days after birth. Then, these cells become nondividing cells and are located in the resting phase of the cell cycle, where their number reduces to 300,000 at puberty. These are present in three types of follicles such as immature, nearly mature, and mature. Generally, mutations in these reproductive cells have lesser frequency than in somatic cells. If this is the case, most of them become lethal during embryonic development.

7.1.2 Spontaneous Mutations

Mutations that take place naturally without human interference, mostly due to the complicated mechanism of gene replication, are referred to as spontaneous mutations. Their frequency is approximately 10^{-5} per gene per generation, which means that the possibility of spontaneous mutation is 1 in 100,000. Generally, nature has a maximum probability of generating the required correct base sequence to form a specific gene. However, there is a minimum probability of creating a wrong sequence, which leads to spontaneous mutations. It may be considered to be either good or bad depending upon the requirement. If the mutation is useful to strengthen the organism in terms of a long, peaceful, and comfortable life, it is considered to be good. Nevertheless, its reverse expression is said to be bad. The human being is also an evolution of several spontaneous mutations that have occurred over millions of years. These are possible due to the availability of several varieties in each gene. These varieties create a "gene pool" with certain flexibility. Whenever, there is a strict necessity to adapt to the environment, such as change in climate, food supply, or surrounding people, the sequence may be modified or a new sequence is created naturally. Change in skin color/pigmentation, change in character, and structural and functional modification of organs are a few examples. In a reverse manner, if there is no usage of that particular sequence for a long time, it may also be eradicated or suppressed, for example, the tail of the monkey.

7.1.3 Genetic Load

Certain mutations produce undesirable characteristics (referred to as bad mutations) such as nondevelopment of hand and feet, bleeding disorders when the blood cannot clot, and malfunctions of critical organs. The deleterious gene responsible for producing bad mutations may be dominant or recessive. The dominant gene expresses its character immediately even though its counterpart on the other chromosome of the pair is normal. If the harmful effect of the dominant gene is so severe, the fertilized ovum cannot develop normally, and this may result in (1) increase in frequency of miscarriages, (2) increase in frequency of still births (neonatal death), (3) reduction in fertility or sterility, and (4) disturbance in the ratio of sexes at birth. This kind of mutation that brings about death before the responsible gene passes its character to the next generation is referred to as "lethal mutation." Hence, the dominant lethal gene is removed in the same generation itself. However, the recessive gene cannot express its character if its counterpart on the other chromosome of the pair is normal. In such a situation, the individual will become a carrier of that particular gene. But a deleterious recessive gene can also lead to lethal mutation if both the parents are carriers of the recessive lethal gene.

It is known that the dominant deleterious gene with lesser harmful effect and the recessive deleterious gene with the most severe (lethal) or lesser effect will pass on to many generations until natural selection causes the gene to vanish from the genome. These deleterious genes constitute the "genetic load (or genetic burden)." The genetic load can be avoided if there is no mutation. In such a case, there is no gene pool and no flexibility for the organism to adapt to environmental changes. Hence, the genetic load (risk) is the price that needs to be paid to have the gene pool (benefit). The size of a genetic load depends on (1) the rate of production of deleterious genes through mutation and (2) the rate of removal of these deleterious genes by natural selection (through failure of cell survival or reproduction). When the rate of production is equal to the rate of removal, the "genetic equilibrium" condition is attained and hence the genome becomes stable over the generations. It can be disturbed by many factors such as mutations, nonrandom mating, small population, immigration, and natural selection.

7.1.4 Mutation Rate

Mutation rate is defined as a measure of the rate at which various types of mutations occur over time. It is difficult to record the formation of gene mutation in human beings due to their long life, few children, uncontrolled mating habits, and so on. So experiments were conducted with smaller organisms that have short life, many progeny per generation, and the possibility to mate under fixed conditions. As they have lesser number of chromosomes than human beings, the site of mutation can be identified easily. These experiments follow an assumption that the system of mutation is the same among all organisms, which has been approved by many scientists. As few scientists refuse to accept this, studies need caution while extending the lower species data to human beings.

The American geneticist H.J. Muller developed a technique to study the probability of occurrence of lethal mutation on chromosomes of fruit flies (*Drosophila* have four pairs of chromosomes). He observed a lethal gene (mutated gene) in a chromosome 1 in 200 times of that chromosomal replication. Since that particular chromosome consists of more than 500 genes and each one can undergo mutation equally, the chances of any one gene in a chromosome undergoing mutation is 1 in 200 × 500 or 1 in 100,000. Studies estimated that the human genome consists of more than 10,000 different genes (discrepancies are there). So the chance of getting a deleterious gene in a single reproductive cell of humans may be 10,000 in 100,000 or 1 in 10. It is also estimated that the possibility for weekly deleterious mutation is four times higher than strongly deleterious mutation (lethal). Then, the chance of getting a weekly deleterious gene in a single reproductive cell may be 4 + 1 in 10 or 1 in 2 (Asimov and Dobzhansky, 1966). Since half of the reproductive cells are free from deleterious genes, there is greater possibility of producing normal generation naturally.

Several studies recorded that the average mutation rate in humans for base substitution, small insertions/deletions, and large structural changes is $1.1–1.7 \times 10^{-8}$ per nucleotide site per generation, ~8% of the base substitution rate and ~0.08 per haploid genome per generation respectively. Furthermore, recent studies have proved that the mutation rate is not a constant. It increases from one generation to the next by a factor 2 in males of age 20–40 due to variation in the rate of cell division. It is also proved to be variable on a geographic scale as the point mutation rate in Europeans is 1.6 times greater than in the Asian population for unknown reasons (Lynch, 2016).

7.1.5 Radiation-Induced Genetic Mutation (Genetic Diseases)

The effect of high-energy X-radiation on the genetic mechanism was first demonstrated by Muller in *Drosophila*. Later, several studies were conducted on offspring of the Japanese atomic bomb survivors, persons exposed for diagnostic and therapeutic purposes, radiation workers, and people living in high background-radiation areas. A study on 70,000 children of Japanese atomic bomb survivors whose parents received an average gonad dose of approximately 15 cGy reports that radiation does not produce any new mutation in a generation but it increases the frequency of spontaneous mutations. Even though there are extensive animal data to prove the genetic effects of radiation, it is not yet confirmed on humans. BEIR VII recorded that the total radiation-induced risk to all types of genetic diseases is in the range of 3.0×10^{-5} to 4.7×10^{-5} rem (3.0×10^{-3} to 4.7×10^{-3} Sv) per year per generation. This means that the possibility of radiation-induced genetic effects is in the range of 3 in 100,000 to 4.7 in 100,000, which is relatively low. The expected genetic effects on the population can be assessed by the "genetically significant dose (GSD)." This is the dose received by the parental gonads during a reproductive life cycle, which starts from conception of the mother to the average age of child-birthing (approximately 30 years in humans).

Furthermore, several studies have shown that the number of mutations is directly proportional to the radiation dose, which gives a straight line graph between dose in the x-axis and mutation in the y-axis. The straight line is continuous without any deviation at low-dose level and crosses the y-axis little above the origin. It confirms that there is no "threshold" for genetic effects of radiation. It also indicates that there is no dose that can be considered as safe. As mentioned earlier, the probability of occurrence of genetic effect increases as the dose increases. Moreover, the mutation (mutation rate) is more than zero even if the dose is zero due to the presence of natural background radiation. Based on this, it is determined that the background radiation contributes to approximately 1% of the spontaneous mutation.

The frequency of radiation-induced mutation is modified by both physical and biological factors. Physical factors include total dose, dose rate, fractionation, and linear energy transfer (LET). Biological factors include germ cell stage, sex, and so on. These have been confirmed through experiments conducted in mice using the specific locus (location of a specific gene in the chromosome) method. This method can detect wide varieties of defects ranging from small intralocus changes to deletion of a few loci. As given by Russel (Russell, 1963), the dose rate has an impact on mutation frequency and depends on the sex and germ cell stage. In males, the effect of dose rate has been confirmed on spermatogonial stem cells (moderate effect) but no effect has been observed on spermatozoa. At low dose rates in the range of 90 R/min to 0.8 R/min, the mutation frequency is approximately 30% of what it is at high dose rates. Hence, it is not increased at higher dose rate (>1000 R/min) or decreased at lower dose rate (<0.001 R/min). In females, mature and maturing oocytes showed a reduction in mutation frequency up to the spontaneous level over the range of dose rates from 1000 to 0.009 R/min. At low dose rates, the mutation frequency is extremely low. But immature oocytes are ineffective even at high doses of low dose rate gamma irradiation, high doses of acute X-rays, and also neutron irradiation. Regarding fractionation, high total doses delivered at high dose rate in fractions at appropriate intervals showed lesser genetic effect in both males and females than the same dose delivered in a single fraction. It also showed that the magnitude of its reduction is the same as the low dose rate effect. At a given dose and dose rate, high-LET radiations induce more mutations than low-LET radiations do.

7.2 Genetic Diseases in Humans

Genetic diseases are classified into three major categories: Mendelian diseases, chromosomal diseases, and multifactorial diseases.

7.2.1 Mendelian Diseases

Mendelian diseases follow the laws proposed by Gregor Johann Mendal in 1865, 1866, and rediscovered in 1900. G.J. Mendal was the German-speaking Augustinian monk who founded genetics and proposed the following three laws:

1. *Law of segregation*—During the formation of gametes, the alleles for each gene segregate from each other so that each gamete carries only one allele for each gene. Gametes are the cells that bind together during sexual reproduction. An allele is an alternative form of one gene.
2. *Law of independent assortment*—During the formation of gametes, genes for different traits can segregate independently. Trait means any single genetically determined characteristic of an organism.
3. *Law of dominance*—Alleles/genes are dominant or recessive; an organism with at least one dominant allele/gene will express the character of the dominant allele/gene.

Mendelian diseases are developed by mutations in a single gene of either autosomes or sex chromosomes. The mutation may be alternation in base, deletion, substitution, insertion, and so on, which may modify either the structure of DNA or the genome. The general point regarding Mendelian diseases is that the relationship between the mutation and the disease is simple and predictable. As mentioned by Eric J. Halls and others, these diseases are classified into autosomal-dominant disorders, autosomal-recessive disorders, and X-linked disorders depending on which chromosome contains the mutated gene and the sequence of transmission.

The dominant disorders follow the law of dominance. In autosomal-dominant disorders, clinically significant abnormality is observed in the first generation itself even any one of the parent's genes in the autosome is mutated even though the other parent has a normal gene. Here, the dominant gene is in the heterozygous state, which will be transmitted to 25%–50% of the progeny of the affected individual. Reports mention that more than 700 autosomal-dominant disorders have been identified. A few examples of autosomal-dominant disorders that appear in adult stage are Huntington's chorea (mental and physical deterioration due to brain disorder), polycystic kidney disease, and multiple polyposis (characterized by cancer of the colon and rectum). Disorders due to congenital abnormalities that appear either at birth or in childhood are polydactyly (formation of extra fingers and toes), achondroplasia (a form of dwarfism due to improper development of cartilage, for example, at the ends of the long bones), brachydactyly (short fingers), and syndactyly (fused fingers).

In autosomal-recessive disorders, clinically significant abnormality is observed only when both of the parent's genes in the autosome are mutated. Here, the recessive gene is in the homozygous state. If one of the parent's recessive genes is mutated, it will not be expressed. Hence, the mutated gene takes many generations to show the autosomal-recessive disorder. Since both of the genes are mutated, most of these disorders are observed at birth or in childhood. A few examples among more than 500 autosomal-recessive disorders are cystic fibrosis (lung disorder that affects the pancreas and other organs), phenylketonuria (metabolic disorder due to faulty phenylalanine amino acid that is caused by mutation in its respective gene), lactose intolerance (inability to digest lactose), adrenal hyperplasia (adrenal gland disorders due to its overgrowth), and sickle cell anemia (blood disorder caused by abnormal form of hemoglobin).

X-linked disorders are caused by mutations in genes located on the X chromosome. Since males have only one X chromosome, only one mutation can induce the diseases that express like dominant mutations. But females have two X chromosomes and hence they require two mutated genes in both of the chromosomes to show its effect as an X-linked recessive mutation. A few examples among more than 80 X-linked disorders are hemophilia (blood fails to clot normally), color blindness, and a severe form of muscular dystrophy (progressive weakness and degeneration of muscles).

7.2.2 Chromosomal Diseases

Chromosomal diseases are induced by either an abnormal number of chromosomes or structural aberrations such as translocations, inversions, breaks, and extra or missing chromosomes formed during meiosis, as discussed in Chapter 3. Studies estimated that about 40% of spontaneous abortions and 6% of fetal deaths are related to chromosomal changes. A few examples among the numerical chromosomal diseases are Down syndrome (three copies of chromosomes in the 21st chromosome), Turner syndrome (presence of only one sex chromosome X), and Klinefelter syndrome (with XXY chromosomes). This kind of numerical chromosomal diseases is not initiated by radiation at low doses, although radiation can cause structural aberrations. A few examples of structural aberrations are cri du chat syndrome (infants produce a high-pitched catlike cry) due to deletion in 5th chromosome and Robertsonian translocations due to fusion of two chromosomes (infants appear normal).

7.2.3 Multifactorial Diseases

The term multifactorial refers to the fact that these diseases have multiple genetic and environmental factors in their origination and transmission pattern. So they are not as simple as Mendelian diseases. Examples of multifactorial diseases include common congenital abnormalities such as neural tube defects, cleft lip with or without cleft palate, congenital heart defects that are present at birth, and chronic diseases of adults such as coronary heart disease, hypertension, and diabetes. One of the environmental factors that increases the risk of heart diseases is intake of high-caloric fatty food. The severity of these diseases varies among persons, families, and populations. The genetic basis of multifactorial diseases comes from family and twin studies. These studies prove that a person has two to six times higher risk than their matched control of getting coronary heart disease if first-degree relatives (parents) are recognized as having the disease and the risk is much higher for twins.

7.3 Genetic Risk Estimation

The goal of genetic risk estimation is to predict the additional risk of genetic diseases in the human population induced by ionizing radiation in addition to natural spontaneous mutations. The concepts of "radiation-inducible genetic diseases" were evolved based on two basic facts and an inference. The facts are that (1) genetic diseases result from mutations that occur in germ cells and (2) ionizing radiation can induce similar changes among all experimental systems. The inference from these facts is that irradiation of human germ cells can increase the frequency of occurrence of genetic diseases in the human population. However, there are no data in humans to prove this inference but it has been proved in other organisms. There are three available methods of estimating radiation-induced genetic diseases: (1) doubling dose method, (2) direct method, and (3) gene number method.

7.3.1 Doubling Dose Method

The doubling dose method is the most commonly used indirect and relatively low-risk method. It expresses the expected increase in mutation frequency per unit of radiation dose in terms of the reference baseline frequency of the mutation. The doubling dose is defined as the dose required to induce the same number of mutations in a generation as arise spontaneously. Hence, it is the ratio between average spontaneous-mutation rate and average induced-mutation rate in a given set of genes. For example, if the average spontaneous-mutation rate is 7.5×10^{-6} per locus in male mice and their average induced-mutation rate is 2.5×10^{-7} per locus, it gives a doubling dose of around 30 R (~0.3 Gy).

$$\text{DD} = \frac{\text{Average spontaneous-mutation rate}}{\text{Average induced-mutation rate}} \tag{7.1}$$

The reciprocal of the DD (i.e., 1/DD) is referred to as the relative mutation risk (RMR) per unit dose. Since DD and RMR are inversely proportional to each other, the smaller the DD, the higher is the RMR and vice versa. Using this DD method, the risk per unit dose is estimated as the product of baseline disease frequency (prevalence of spontaneously arising genetic diseases), P and 1/DD (RMR). That is,

$$\text{Risk per unit dose} = \text{P} \times \left(\frac{1}{\text{DD}} \right) \tag{7.2}$$

The population genetic theory also insists on the use of Equation 7.2 to explain the genetic equilibrium in the population. The theory assumes that there is stability in the frequency of genetic mutations. This is due to the balance between the rate at which spontaneous mutations enter the gene pool in every generation and the rate at which they are eliminated by natural selection.

When the mutation rate is increased due to exposure to ionizing radiation in every generation, the genetic equilibrium is disturbed so that the population will reach a new equilibrium condition between mutation and selection. The amount of increase in mutation frequency, the time it takes for the population to reach this new equilibrium condition, and the rate of approach to equilibrium depend on the induced-mutation rate, its selection possibility, the type of genetic disease, and the number of generations exposed. Hence, the risk per unit dose in a population that attains new equilibrium is increased since its initial population is assumed to be in genetic equilibrium condition.

7.3.2 Direct Method

The direct method is used to estimate the absolute probability of occurrence of dominant diseases induced by radiation in the first-generation progeny. In this method, the induction rate for a specific type of defects, such as cataract and skeletal anomalies, is measured directly in mice irradiated with high–dose rate radiation. Then, the proportion of serious dominant diseases in humans that involve similar defects is estimated. This is used as a proportionality factor to estimate the effect of radiation on all dominant mutations in humans. For example, if the spermatogonial chronic induction rate for skeletal defects in mice is 7.5×10^{-6} rad/gamete and if about one in five serious dominant disorders in humans involved the skeleton, then the first-generation effect of spermatogonial chronic irradiation estimated by this direct method is 20 induced cases/10^6 live births/rad. Since this method does not rely upon knowledge of the natural prevalence of genetic diseases in populations, it requires further analysis by considering the variation in radiosensitivity and immunity among different species, germ cell stage, relative viability of chromosomal aberrations, transmission rates, and so on.

7.3.3 Gene Number Method

The gene number method is an old method that was used to estimate the total number of radiation-induced mutations using the following equation:

$$\text{Number of induced mutations} = \text{Number of genes} \times (\text{induction rate / gene / dose}) \times \text{dose} \tag{7.3}$$

As there is no satisfactory method to estimate the total number of mutable genes in humans and it is hard to translate to human population, this method is not used for risk estimation.

7.4 Background Data from Humans and Other Animals

Initially, Muller conducted experiments in fruit flies and estimated the doubling dose in the range of 5–150 R (0.05–1.5 Gy for X-rays) for mutations. After the Second World War, Russells and coworkers carried out the "megamouse project" with 7 million mice to determine the specific locus mutation under various physical and biological conditions such as sex, dose, dose rate, and fractionation. In these experiments, seven specific locus mutations (six for coat color and one for stunted ear) were used to estimate the doubling dose of 1 Gy for humans. They inferred that the rate of mutation is significantly reduced when a time gap of 2 months in male mice and more in female mice is maintained between irradiation and insemination. The same data are extended to humans with certain assumptions, and a period of 6 months is recommended. Based on this, it is advised to have a 6-month gap between irradiation and a planned conception when a person accidentally, occupationally, or from diagnostic and therapeutic procedure receives a considerable amount of gonadal dose (~0.1 Gy).

In order to estimate the genetic risk in the human population, the baseline spontaneous mutation rate and the doubling dose calculated from mice experiments are required. It also requires two correction factors. One correction factor is to allow for the fact that not all mutations lead to a disease. This is represented by a mutation component (MC). The MC varies among various kinds of genetic diseases. For the first-generation effect of autosomal-dominant and X-linked diseases, for autosomal-recessive diseases, and for chronic multifactorial diseases, it is 0.3, 0, and 0.01–0.02, respectively. Another correction factor is to allow for the fact that the seven specific locus mutations used to estimate the doubling dose in the megamouse project are not representative of the collection of radiation-induced genetic diseases.

The UNSCEAR 2001 estimated the genetic risk of autosomal-dominant and X-linked diseases for the first generation at around 750–1500 cases/million live births/gray of chronic low-LET radiation. This is compared with the baseline spontaneous-mutation risk of 16,500 cases/million. But the genetic risk of autosomal-recessive diseases is 0 compared with the baseline spontaneous mutation risk of 7500 cases/million. The genetic risk of chronic diseases is about 250–1200 cases/million compared with the baseline spontaneous mutation risk of 650,000 cases/million. The genetic risk of multifactorial diseases and congenital disorders is around 2000 cases/million population. However, the total risk/gray is just 0.41%–0.64% of the baseline spontaneous-mutation risk of 738,000 cases/million live births.

The International Commission of Radiological Protection (ICRP) also estimated the genetic risk of radiation at 0.2% per sievert and 0.1% per sievert for public and occupational radiation, respectively. Based on the studies conducted on more than 72,000 children of the atomic bomb survivors, a doubling dose in the range of 1–2 Sv has been estimated. These studies show that the importance of genetic effects of radiation are reduced continuously but the low-dose effect of radiation that may cause carcinogenesis is a major concern in this era.

Objective-Type Questions

1. The genetic effects of X-rays were initially identified by
   a. Muller
   b. Russel
   c. Mendal
   d. None
2. Mutation can be induced by
   a. Chemical
   b. Radiation
   c. Environment factor
   d. All
3. Polyploidy is a
   a. Gene mutation
   b. Chromosome mutation
   c. Somatic mutation
   d. Genetic mutation
4. The possibility of spontaneous mutation is
   a. 1 in 100
   b. 1 in 1,000
   c. 1 in 10,000
   d. 1 in 100,000
5. Three copies of chromosomes in the 21st chromosome is referred to as
   a. Down syndrome
   b. Turner syndrome
   c. Klinefelter syndrome
   d. Robertsonian translocations
6. The doubling dose estimated from the atomic bomb survivors is
   a. 0.5–1 Sv
   b. 1–2 Sv
   c. 0.5–1 mSv
   d. 1–2 mSv
7. The following is an autosomal-recessive disorder
   a. Polydactyly
   b. Achondroplasia
   c. Brachydactyly
   d. Adrenal hyperplasia

8. The expected genetic effect on the population can be assessed by the
   a. Genetic load
   b. Genetically significant dose
   c. Genetic risk
   d. All
9. The reciprocal of the doubling dose is referred to as
   a. Relative mutation risk
   b. Relative risk factor
   c. Genetic risk
   d. Risk per unit dose
10. The mutation component for chronic multifactorial disease is
    a. 0
    b. 0.01
    c. 0.02
    d. 0.3

8

Radiobiological Models

Objective

This chapter highlights the various radiobiological models available to understand the effects of radiation on cells, their principles, and limitations for further development.

8.1 Importance of Radiobiological Models

Radiobiological models (RBMs) are widely used in both conventional and modern radiotherapy for various reasons as discussed here. In conventional X-ray radiotherapy, RBMs are used to obtain a new dose prescription (or the isoeffective doses) whenever the dose prescription for the standard/ongoing fractionation is altered by considering (1) the normal tissue sensitivity, (2) varying the number of fraction, or (3) shrinkage in tumor volume. But, in modern radiotherapy, RBMs are used to convert some physical quantities such as absorbed dose to biological quantities such as biologically equivalent dose (BED), tumor control probability (TCP), and normal-tissue complication probability (NTCP). These biological quantities are required to optimize the treatment plan by using them in biologically based treatment-planning systems (TPSs). However, in proton or carbon ion beam therapy, RBMs are proposed to obtain the relative biological effectiveness (RBE) for various beam conditions in order to compare the clinical data with the reference beam (X-ray). Different researchers, namely Strandquist (1944), Lea (1946), Ellis (1967), Orton (1973) among others, have developed various RBMs based on their knowledge of the mechanism that determines the cellular response to radiation. The development of radiobiological concepts and their related physical parameters that are incorporated in these models, as described by Hall and Giaccia (2006), Wang (2010), Garzón et al. (2014), and Bodgi et al. (2016), are discussed in this chapter.

8.2 Models Based on Empirical Isoeffect

In 1944, Strandqvist developed a novel empirical (derived from experiment and observation rather than theory) approach to analyze the biological effects of radiation by introducing the dose isoeffect through the isoeffective curve. He performed experiments to analyze

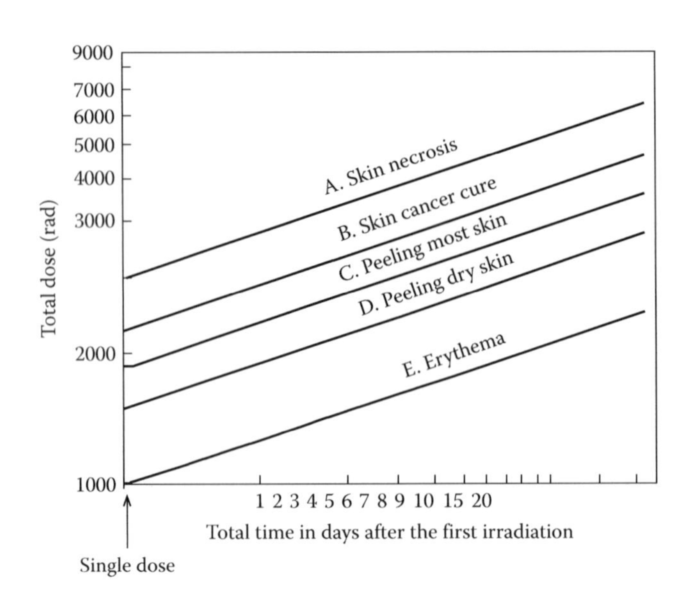

FIGURE 8.1
Strandqvist isoeffect curve. (From Strandquist, M., *Acta Radiol.*, 55(suppl.), 1–300, Copyright 1944, with permission from Sage publications.)

the control of cancer effects on skin using a conventional scheme of 2 Gy/day and 5 days/week in a 250 kV_p X-ray machine. From this, he had drawn the isoeffect curve as shown in Figure 8.1. The isoeffect curve relates the variation in total dose of the fractionated treatment and the total treatment time (in days) to produce an equivalent bioeffect (skin reactions such as erythema, peeling dry skin, peeling moist skin, skin cancer cure, and skin necrosis).

The isoeffect curve of the time–dose model was expressed as follows:

$$D = kT^{1-p} \tag{8.1}$$

where
- D is the total dose
- T is the treatment time
- p is a potency to be determined which is <1
- k is the constant of proportionality

The isoeffect curve was further analyzed by Cohen in order to compare the response data between normal and cancerous tissues. The exponent of T (i.e., 1 − p) in the equation is equal to the slope of the isoeffect curves. It represents the average value of the repair and recovery capacities of normal or cancerous tissues. Knowing that the slope of the curve corresponds to normal tissue as 0.33 and cancerous tissue as 0.22, the relationship between the tolerance dose and total treatment time was derived in order to cure the skin tumor with minimal complications as expressed in the following equations:

$$Dn = k_1T^{0.33} \quad \text{for skin erythema/skin tolerance} \tag{8.2}$$

$$D_t = k_2T^{0.22} \quad \text{for skin cancer} \tag{8.3}$$

8.2.1 Nominal Standard Dose Model

In continuation with Cohen, Fowler and Stern carried out empirical research on pig skin. They indicated that the repair capacity of normal tissue was larger than that of cancerous tissue, that homeostatic recovery took longer than intracellular recovery, and confirmed that the isoeffect dose varied not only with treatment time but also with the number of fractions. Based on this, Ellis proposed a mathematical expression relating the total dose, treatment time, and the number of fractions. He assumed that the exponent of T for normal tissue consisted of two parts: (1) the intracellular recovery represented by exponent 0.22 and (2) the homeostatic recovery represented by exponent (0.33 – 0.22) = 0.11. Hence, the isoeffect Equation 8.2 was modified as follows:

$$D = k_1 T^{0.22} T^{0.11} \tag{8.4}$$

Then, he replaced the term $T^{0.22}$ into n^p, where n is the number of fractions, which again modified Equation 8.4 as follows:

$$D = k_1 n^p T^{0.11} \tag{8.5}$$

Further, Ellis used a conventional scheme of 30 fractions of 2 Gy/fraction delivered over 40 days to find the value of $\mathbf{n^p}$ using the following relationship:

$$40^{0.22} = 30 \quad \text{which gives } p = 0.24$$

So, Equation 8.5 was modified into

$$D = k_1 n^{0.24} T^{0.11} \tag{8.6}$$

He adjusted the constant of proportionality k to a treatment scheme and called it nominal standard dose (NSD), which consisted of dose and time parameters with the unit of rad-equivalent therapy (RET). Then, Equation 8.6 became

$$D = (NSD) n^{0.24} T^{0.11} \tag{8.7}$$

(or)

$$NSD = D n^{-0.24} T^{-0.11}$$

Even though the NSD concept was well received to calculate the tolerance dose of normal tissue in the skin cancer region and its connective tissues, it has the following limitations:

1. It cannot be applied to cancers and normal tissues other than skin and its connective tissues.
2. It cannot be suited to therapy interruptions and does not incorporate volume effect.
3. It is not suitable in treatments where less than 4 fractions (including single equivalent dose) are selected since Ellis used data from 4 to 30 fractions.

8.2.2 Time–Dose Fractionation Factor Model

In order to rectify the limitations in the NDS concept, Orton and Ellis introduced the time–dose fractionation (TDF) factor in the NSD Equation 8.7 and formed a new model called the TDF model. This model relates the biological effect of two therapeutic schemes using TDF value if the biological effects and TDF value of those two schemes are the same. They modified Equation 8.7 by replacing D with nd, where n is the number of fractions and d is the dose/fraction and T with T/n for a fixed number of fractions:

$$\mathrm{NSD} = \mathrm{nd}\,\mathrm{n}^{-0.24}\left(\frac{\mathrm{T}}{\mathrm{n}}\right)^{-0.11} \tag{8.8}$$

(or)

$$\mathrm{NSD} = \mathrm{d}\left(\frac{\mathrm{T}}{\mathrm{n}}\right)^{-0.11} \mathrm{n}^{0.65}$$

Raising both sides of Equation 8.8 to the power of 1.538 and multiplying with a scaling factor of $\mathbf{10^{-3}}$, they obtained

$$\mathrm{TDF} = 10^{-3}\mathrm{NSD}^{1.538} = \mathrm{nd}^{1.538}\left(\frac{\mathrm{T}}{\mathrm{n}}\right)^{-0.17} 10^{-3} \tag{8.9}$$

In the case of treatment with interruption, TDF after the break (gap correction) is calculated by reducing the TDF before the break with the following decay factor:

$$\text{Decay factor} = = \left[\frac{\mathrm{T}}{(\mathrm{T}+\mathrm{R})}\right]^{0.11} \tag{8.10}$$

where
T is the treatment time in days from the beginning of the treatment
R is the rest of the days

(or)

$$\text{Decay factor} == \left[\frac{\mathrm{T}}{\mathrm{T}+\mathrm{G}}\right]^{0.17}$$

where
T is the duration of treatment in days
G is the gap in days

The TDF model has the following advantages:

1. The linear behavior of the TDF with respect to the total number of fractions N.
2. It is easy to apply clinically.
3. It relates the basic variables of radiotherapy such as dose, time, and number of fractions.

However, the TDF model needs to be improved to overcome its short range of applicability on various tissues.

8.3 Models Based on Cell Survival Curves and Isoeffect

8.3.1 Cell Survival Curve

A survival curve of mammalian cells is used to get direct information on their response to radiation. It is drawn between the absorbed dose on a linear scale on the x-axis and the natural logarithm of the surviving fraction of cells on the y-axis. Two methods, namely in vitro and in vivo, are used to determine the survival fraction (SF) of cells when they are exposed to a range of doses.

In vivo methods (live study) are based on analyzing the reduction in the number of cells due to irradiation to an organ of the same animal or another animal. It can be studied by a clonogenic assay. These methods are used to study the relation between the cell survival and the dose in both normal and cancerous tissues. The commonly used in vivo clonogenic assays are the exogenous colony assay and the endogenous colony assay. In the exogenous colony assay, cells (hemopoietic, mammary, and thyroid cells) from the irradiated animal are transplanted into another animal where the survived cells are allowed to divide and form colonies. Whereas, in the case of endogenous colony assay, cells (small intestine, testis, hemopoietic, skin, and kidney cells) from the irradiated animal are left undisturbed in the same animal to divide and form colonies for further analysis.

In vitro methods are based on the ability of surviving cells to reproduce/divide in the culture medium (outside the organism) after irradiation. To determine the SF of cells by the in vitro method, cells from an actively growing stock culture are mixed with the enzyme trypsin. It dissolves and loosens the cell membrane to make a single-cell suspension. The number of cells per unit volume in this suspension is then counted using a hemocytometer or a Coulter counter. From this suspension, a known number of cells, for example 100 cells, are seeded into a culture dish without irradiation and the dish is incubated for approximately 12–14 days depending on the division rate of the cell. This is considered as a control sample for this procedure. Each single cell in the culture dish divides many times and forms a colony of its progeny. These colonies are easily visible to the naked eye.

As shown in Figure 8.2, for the 100 cells (figure denoted for just 10 cells to simplify it) seeded into the dish, the number of colonies counted may be from 0 to 100. One may expect the maximum number of colonies to be 100. But this is not possible for several reasons, including suboptimal growth medium, errors in counting the number of cells initially plated, and the loss of cells by trypsinization and general handling. From the number of colonies counted from the unexposed population of cells (control sample), the "plating efficiency (PE)" is calculated using the following formula:

$$\text{Plating efficiency (PE)} = \frac{\text{Number of colonies counted}}{\text{Number of cells seeded}} \times 100$$

Plating efficiency indicates the percentage of cells seeded into a culture dish that finally grows to form a colony. If 50 colonies are counted out of 100 cells, then the plating efficiency is 50%. As a parallel step, a known numbers of cells which are irradiated to known doses (2, 4, 6 Gy etc.) of gamma rays are allowed to divide and form colonies in a number

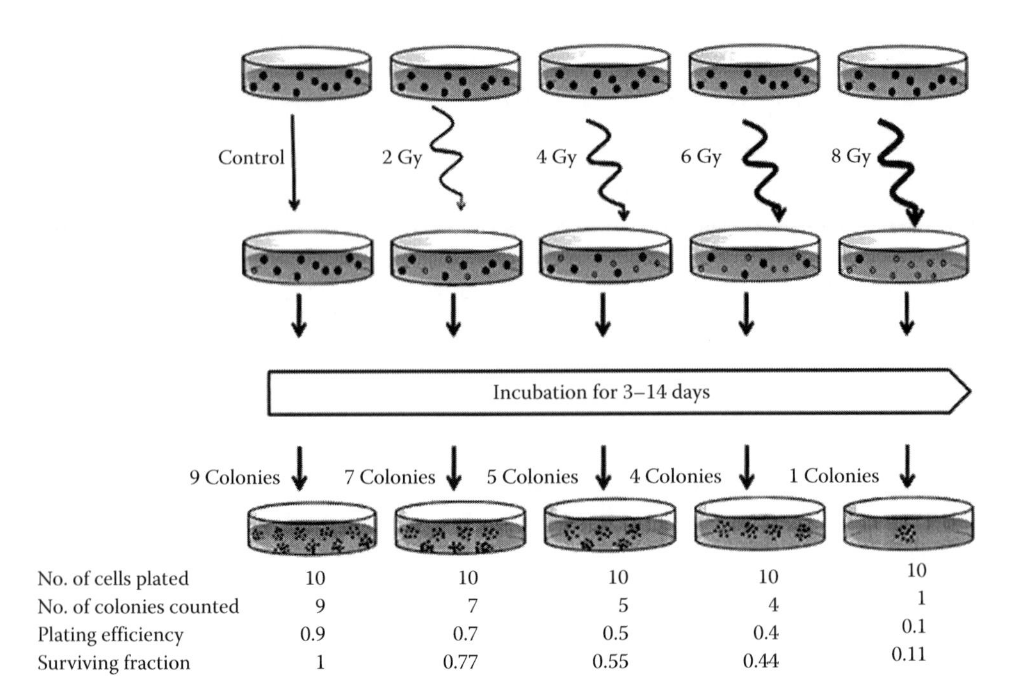

FIGURE 8.2
Schematic representation of the steps involved in in vitro cell survival curve.

of culture dishes like the control sample. When we analyze the number of colonies in each dish, the following may be noticed:

1. Some cells may remain single, and not divide. These cells are scored as dead.
2. Some cells may go through one or two divisions and form small colonies of just a few cells. These cells are also scored as dead.
3. Some cells may form large colonies. This is due to the presence of surviving cells, which retain their dividing capacity after irradiation. These cells are scored as alive.

Based on these, the surviving fraction is calculated by

$$\text{Surviving fraction (SF)} = \frac{\text{PE of irradiated sample}}{\text{PE control sample}} \times 100$$

Using this method, various types of survival curves like linear/exponential and curved and shouldered survival curves are obtained based on different models in order to get a curve that is consistent with the experimental data. Further, the shapes of these curves are also varied depending on the type of radiation (low and high linear energy transfer (LET) radiation), tissues of various radiosensitivity (early responding, late responding normal tissues and cancerous tissues), dose rate, oxygen effect, fractionation, and so on, as discussed in Chapter 4.

8.3.2 Random Nature of Cell Killing and Poisson Statistics

It is well known that the nature of energy deposition along the tracks of ionizing radiation within the cell is a random process and the inherent nature of the biological system (genetic makeup for DNA repair, tumor/tissue microenvironment, etc.) is also unpredictable. Hence, the nature of cell killing/survival also follows random probability statistics and uncertainties. Considering these factors, a Poisson-based model was developed to understand the shape of the cell survival curve in a simple way by assuming that the mean number of cells killed is proportional to the dose. In other words, the probability of survival is decreased with a dose that shows a simple exponential survival curve with a slope. It can also be represented by the following Poisson statistics to find the probability of a specific number of event/hit/ionization (represented by H here) where x is the average number of events:

$$P(H) = \frac{\left[\left(e^{-x}\right)\left(X^{H}\right)\right]}{H!} \tag{8.11}$$

If each "hit/ionization" is assumed to cause cell inactivation/killing, the probability for a single cell survival is the probability for a zero hit P(0), where x = 1. It gives the value of D_0, which is referred to as the mean lethal dose, or D_{37} or D_0 is the dose that reduces the cell population to 37% as given in the following equation. Since D_0 or D_{37} is the dose that produces one mean lethal event per target, it is used to represent the slope of the survival curve. Similarly, D_{10} is the dose that reduces the cell survival to 10%. D_0 and D_{10} can be related as $D_{10} = 2.3 \times D_0$.

$$P(0) = \frac{\left[\left(e^{-1}\right)\left(1^{0}\right)\right]}{0!} = e^{-1} = 37\%$$

In continuation to this simple Poisson-based model, scientists derived many mathematical expressions along with their theoretical assumptions and developed a number of survival-based models, such as target theory, linear quadratic model (LQM), and local effect model (LEM). These models are discussed here briefly with their specific advantages and limitations.

8.3.2.1 Target Theory

Target theory was developed by Lea in 1946 using studies on the effects of ionizing radiation on microorganisms and bioactive molecules. He assumed that his model was also directly applicable to radiation-induced chromosomal aberrations. He proposed a list of basic concepts to explain radiation-induced damage that are based on the number of targets within the cells and their volume. In order to understand target theory, let us consider the target as a unit of biological function; hence, inactivation of this target can cause cell death. Based on these principles, Lea developed his target theory by incorporating the following assumptions and limitations:

1. The random energy deposition events that produce the biological damages are active events to produce hits, but the secondary events are not included.
2. The targets present in the cells are simple and of the same size.
3. The total number of events recorded in the total population of cells is the total cell volume multiplied by the dose.
4. One or more hits are required to induce target inactivation.
5. This theory is suitable to study the effects of only low-LET radiations.

8.3.2.2 Single Target–Single Hit Model

The single target–single hit model is also referred to as simple target theory. In this model, it is assumed that each cell consists of a single target and a single hit is enough to inactivate the cell, which leads to cell death. If dN is the number of targets inactivated out of N targets present, then this is proportional to the total number of targets and the total dose D:

$$dN \alpha ND \quad \text{i.e.,} \quad N = -N_0 \frac{dD}{D_0} \tag{8.12}$$

where

D_0 is the mean lethal dose (D_{37})
N_0 is the initial number of targets/cells

The negative sign indicates the reduction in the number of cells due to cell death. D_0 characterizes the intrinsic sensitivity of a cell population. If D_0 is small, the radiosensitivity of the cell is high, that is, a smaller dose is required to reduce the initial cell survival to 37%. By integrating Equation 8.12, we get

$$N = N_0 e^{-D/D_0} \tag{8.13}$$

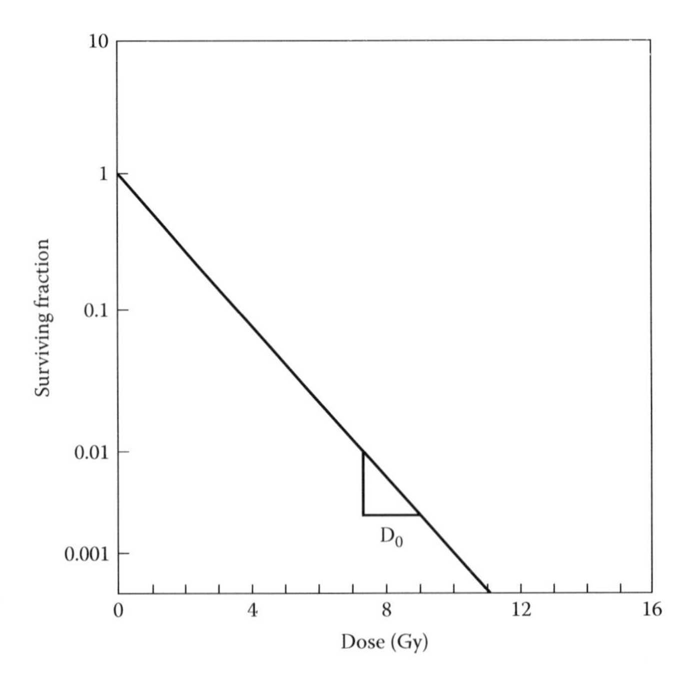

FIGURE 8.3
Schematic representation of the single target–single hit survival curve. (From Kelsey, C.A., Heintz, P.H., Sandoval, D.J., Chambers, G.D., Adolphi, N.L., and Paffett, K.S.: *Radiation Biology of Medical Imaging*, 1st edn. 87. 2014. Figure 5.4. Copyright Wiley-VCH Verlag GmbH & Co. KGaA. Reproduced with permission.)

The cell surviving fraction SF is represented as follows:

$$SF = \frac{N}{N_0} = e^{-D/D_0} \quad \text{i.e., } \ln S = -\left(\frac{D}{D_0}\right) \tag{8.14}$$

The survival curve that corresponds to this model is also exponential, as shown in Figure 8.3. It shows that as the dose increases, the number of surviving cells reduces (or cell killing is increased) exponentially. The limitation of the single target–single hit model is that it is a poor model to study the effects of radiation on mammalian cells as it uses data derived from microorganisms, but it is used to explain the multitarget model and LQM.

8.3.2.3 Multitarget–Single Hit Model

The multitarget–single hit model was used extensively for many years and still has some advantages over other models. The survival curve of mammalian cells exposed to low- and high-LET radiations obtained by this multitarget model is shown in Figure 8.4. In this model, it is assumed that the cell consists of multiple targets and each of these should be hit once to achieve cell death. Even if there is only one target that remains unaffected, this can spare the cell from death. The survival curve for high-LET radiation (alpha particle, neutron, etc.) is a straight line from the origin. This means that the survival fraction is reduced exponentially as the dose of high-LET radiation is increased. This curve can be characterized by a single parameter, the slope D_0. At low doses of low-LET radiations (X- and gamma-rays), sparse ionization is produced. So, the radiation cannot hit all the targets in the cell which shows an

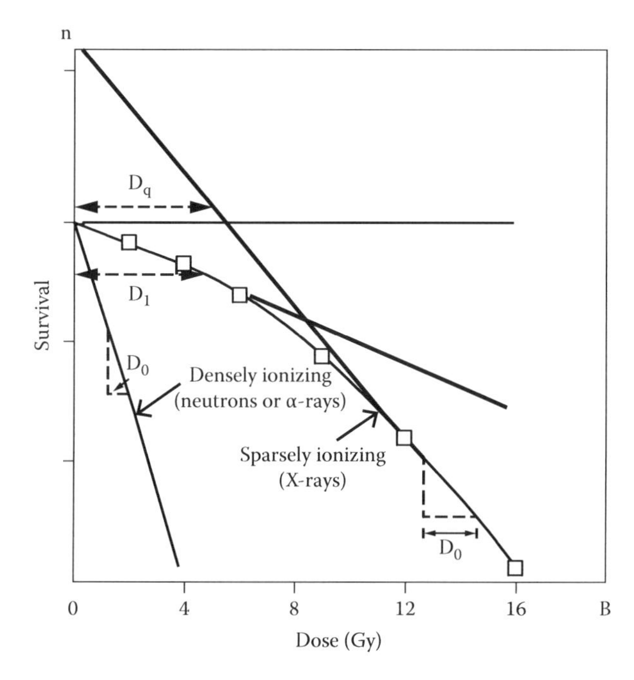

FIGURE 8.4
The multitarget model–based survival curve of mammalian cells after exposure to low- and high-LET radiations. (From Kelsey, C.A., Heintz, P.H., Sandoval, D.J., Chambers, G.D., Adolphi, N.L., and Paffett, K.S.: *Radiation Biology of Medical Imaging*, 1st edn. 91. 2014. Figure 5.7. Copyright Wiley-VCH Verlag GmbH & Co. KGaA. Reproduced with permission.)

initial linear region in the curve. As the dose is increased, the number of targets hit by radiation also increases and at a certain dose, cell death is achieved. By further increasing the dose, the number of cells killed by radiation also increases exponentially, which forms another linear downward region in the curve. According to the previous interpretation and assumptions, the survival equation to relate the surviving fraction (SF) with the dose (D) for this model, which has n number of targets, is represented as

$$SF = 1 - \left(1 - e^{-D/D_0}\right)^n \tag{8.15}$$

The survival curve of the multitarget model can be characterized by D_1, D_0, n, and D_q. D_1 is the reciprocal of the initial slope of the curve at its initial linear portion that results from single-event killing. It is the dose required to reduce the cell surviving fraction from 1 to 0.37. As mentioned earlier, D_0 is the mean lethal dose, which is the reciprocal of the final slope of the curve at the high-dose linear portion that results from multiple-event killing. It is the dose required to reduce the cell surviving fraction from 0.1 to 0.037 or from 0.01 to 0.0037, and so on; n is the extrapolation number of the curve which is the measure of the shoulder width of the curve. This is obtained by extrapolating the linear part (high-dose region) of the curve backward to intersect with the y-axis of the curve. If n is larger (>10), the survival has a broader shoulder and vice versa. D_q is the quasi-threshold dose that is also used to represent the width of the shoulder. As there is no dose below which radiation does not produce any effect, the term quasi-threshold (almost threshold) dose is used instead of threshold dose. It is the dose at which the extrapolated line from the final straight portion of the survival curve cuts across the axis at a surviving fraction of unity. The relationship among the parameters D_0, n, and D_q is expressed by the following equation:

$$\log_e n = \frac{D_q}{D_0} \tag{8.16}$$

The limitations of the multitarget–single hit model are as follows:

1. It does not define DNA as a susceptible target.
2. It does not present the real processes inside the cell such as kinds of damage, induction, and repair of those damages.
3. It does not consider any time-dependent quantities.

Exercise: Draw a survival curve with $D_0 = 3$ Gy, and $n = 5$. From this, find the value of D_q and compare your answer using the relationship between D_0, n, and D_q.

8.3.3 Linear Quadratic Model

The linear quadratic model (LQM) was initially formulated as an empirical model used to fit radiation-induced chromosome aberrations and later incorporated the basis of cell survival curves. It was developed to explain the dose response of exchange type of chromosome aberrations that result from two independent breaks in the chromosomes. It is one of the powerful and widely used models in radiotherapy. This model assumes that radiation-induced cell killing has two components: one proportional to dose, which is a linear component αd, and another proportional to the square of the dose, which is a quadratic component βd^2.

The linear component is represented by the initial slope of the curve, which relates to the irreparable DNA double-strand breaks that lead to chromosome breaks, which in turn are

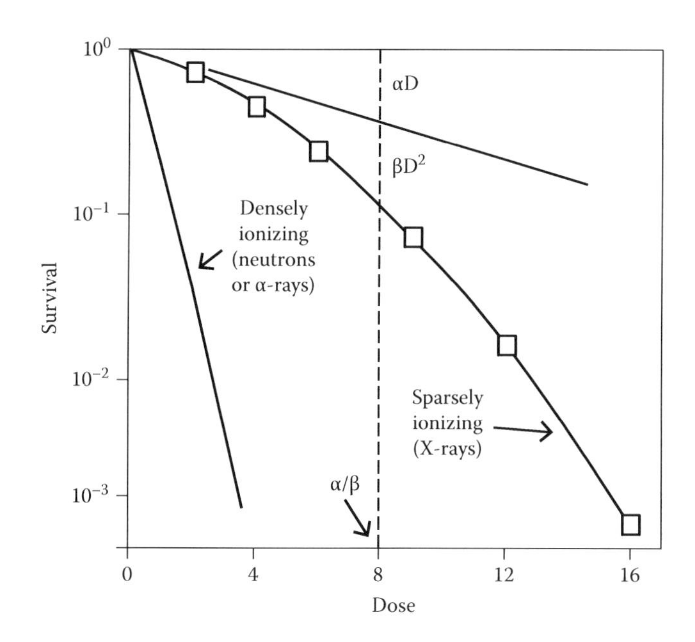

FIGURE 8.5
Linear quadratic model–based cell survival curve. (From Kelsey, C.A., Heintz, P.H., Sandoval, D.J., Chambers, G.D., Adolphi, N.L., and Paffett, K.S.: *Radiation Biology of Medical Imaging*, 1st edn. 91. 2014. Figure 5.7. Copyright Wiley-VCH Verlag GmbH & Co. KGaA. Reproduced with permission.)

lethal to the cell. But the quadratic component is represented by the final slope of the curve. It relates to lethal damage, which can be produced either by a single radiation track (i.e., proportional to αd) or by the interaction of two or more less severe DNA lesions (SLDs) produced by two independent radiation tracks (i.e., proportional to βd^2). The two components of cell killing are equal when $\alpha d = \beta d^2$ or $d = \alpha/\beta$. This means that the linear and quadratic components are equal at a dose that is equal to the ratio of α to β (α/β), as shown in Figure 8.5. The α/β parameter is very important in radiotherapy. It is related to the radiosensitivity of cells and their repair capacity. A high α/β indicates a cell with a high turnover capacity and a low repair capacity.

Figure 8.5 shows that for low-LET radiation at a low dose range, the βd^2 component is insignificant, and the survival curve is proportional to the dose, and hence it is linear. At higher doses, the cell survival is proportional to the square of the dose and the curve is concave downward. In the case of high-LET radiations, β damage is zero since its ionization track induces DNA double-stand breaks, and hence the survival curve is linear.

According to this model, the survival fraction S at a dose per fraction d in conventional fractionated X-ray radiotherapy is expressed as follows:

$$S = e^{-\left(\alpha d + \beta d^2\right)} \tag{8.17}$$

where α and β are initial and final slopes or coefficients of irreparable and repairable damage, respectively.

The survival for n number of fractions can be expressed as

$$S_n = e^{-n\left(\alpha d + \beta d^2\right)} \quad \text{or} \quad -\ln S_n = n\alpha d + n\beta d^2 \tag{8.18}$$

In radiotherapy, the value of the biologically effective dose (BED) is used widely to compare various fractionation schemes. BED is obtained by dividing the survival fraction of cells as follows:

$$\frac{-\ln S_n}{\alpha} = nd + n\frac{\beta}{\alpha}d^2 = nd\left(1 + \frac{\beta}{\alpha}d\right) \tag{8.19}$$

$$\text{BED} = -\frac{\ln S_n}{\alpha} = nd\left(1 + \frac{d}{\alpha/\beta}\right) \tag{8.20}$$

LQM has also been used in the formula for calculating tumor control probability (TCP) and normal-tissue complication probability (NTCP) in modern radiotherapy (given in the next section). However, one should be cautious while using this model as it does not include the factors associated with various patient-specific predictive assays. Moreover, LQM is based on the response of a single cell and hence it does not include the effects produced by the intercellular communications in a tumor and normal tissue microenvironment.

8.3.4 Local Effect Model

Since the LQM is suitable only for use in radiotherapy of conventional low-LET radiations, many modifications have been obtained in order to extend the same for high-LET radiations. Among them, the local effect model (LEM) alone has been implemented clinically for use with proton and carbon ion beams. It is well known that the clinical outcome of ion beam therapy is better than that of photon therapy as it improves TCP with reduced NTCP due to the presence of Bragg peak. In LEM, the cell survival curve of high-LET radiation is obtained by combining the in vitro cell survival data of LQM for low-LET radiations with the track structure of high-LET radiations. In addition, data of small animals are used to determine the RBE (discussed in Chapter 4) values of those radiations. The model assumes that the local biological effect (DNA damage, chromosomal aberrations, etc.) in a small subvolume of the cellular nucleus can be determined by the dose in that subvolume and is independent of the type of radiation. Even though the biological end point and dose delivery scheme are well defined in the LEM, it necessitates further improvement because the value of RBE (discussed in Chapter 4) is subject to uncertainties due to the stochastic nature of energy deposition of radiation tracks and complex processes are involved in DNA damage and repair.

8.4 TCP- and NTCP-Based Models

The most commonly used TCP- and NTCP-based models in TPSs, as given in AAPM Task Group 166, are discussed here briefly.

8.4.1 Common TCP Models

Most of the TCP models are based on the assumption that the number of surviving clonogenic tumor cells follows the Poisson distribution. If the initial number of clonogens is N_c, and S is the overall surviving fraction after treatment, the average number of surviving clonogens is taken as SN_c. Then, the TCP is equal to the probability that no clonogens survive, and is represented as

$$TCP = \exp(-SN_c) \tag{8.21}$$

In commonly used empirical TCP models, TCP can be approximated by D_{50}, and the normalized dose gradient $\gamma = D\frac{dTCP}{dD}$ evaluated at $D = D_{50}$. Generally, TCP is represented as a product over the structure's voxel weight probability functions as follows:

$$TCP = \prod_{i=1}^{M} P(D_i)^{vi} \tag{8.22}$$

where
 M = number of voxels
 v_i = relative volume of the voxel, i.e., $v_i = V_i/V_{ref}$

Based on the LQ model, $P(D_i)$ can be written as

$$P(D_i) = \exp\left(-\exp\left(e\gamma - \alpha D_i - \beta\frac{D_i^2}{n}\right)\right) \tag{8.23}$$

where
$$\alpha = \frac{e\gamma - \ln(\ln 2)}{D_{50}\left(1 + \frac{2}{\alpha/\beta}\right)}$$

$$\beta = \frac{e\gamma - \ln(\ln 2)}{D_{50}\left((\alpha/\beta) + 2\right)}$$

Based on the linear Poisson formulation, $P(D_i)$ can be written as

$$P(D_i) = \exp\left(-\exp\left(e\gamma - \frac{D_i}{D_{50}}\left(e\gamma - \ln(\ln 2)\right)\right)\right) \tag{8.24}$$

8.4.2 Common NTCP Models

The volume effect is the major concern in modeling the dose–response relationship for normal tissues. As mentioned in Chapter 4, an organ may be a parallel organ or a serial organ. In parallel organs, organ function can be protected considerably by reducing the irradiated volume and their response is also well correlated with a mean organ dose. But serial organs show threshold-like responses to radiation that are well correlated with a maximum organ dose or the hot spot. The most commonly used NTCP-based models in treatment planning that can account for volume effect are described here:

1. *Lyman–Kutcher–Burman (LKB) model*: This model was proposed in 1985 to describe complication probabilities for uniformly irradiated whole- or partial-organ volumes. Since normal tissues are rarely irradiated uniformly, Kutcher and Burman designed an effective volume-based model in 1989 in order to convert a

heterogeneous dose distribution into a uniform whole-body irradiation resulting in the same NTCP. The combined formalism is referred to as the LKB model. According to this model, NTCP can be calculated using the following equation:

$$\mathrm{NTCP} = \frac{1}{2\pi}\int_{-\infty}^{t} e^{x^2/2}dx$$

$$t = \frac{D_{\mathrm{eff-TD50}}}{mTD_{50}}$$

$$D_{\mathrm{eff}} = \left(\sum_i v_i D_i^{1/V}\right)^V$$

where

D_{eff} is the dose uniformly given to the entire volume that can lead to the same NTCP as the actual heterogeneous dose distribution

TD_{50} is the uniform dose given to the whole organ that can result in 50% complication risk

m is the measure of the slope of the sigmoid curve

V is the volume effect parameter

v_i is the fraction of organ volume receiving a dose D_i

2. *Relative seriality model*: The relative seriality model, or the s-model, explains the response of an organ which has a combination of both serial and parallel functional subunits (FSUs; refer to Section 4.2.5). According to this model, the NTCP is given by the following equation:

$$\mathrm{NTCP} = \left\{1 - \coprod_{1}\left[1 - P(D_i)^s\right)^{v_i}\right\}^{1/s}$$

where

s is the relative contribution of each type of architecture

v_i is the fraction of organ volume receiving a dose D_i

$P(D_i)$ is the complication

Objective-Type Questions

1. Mean lethal dose is

   a. D_{10}

   b. D_{30}

   c. D_{37}

   d. D_{50}

2. The survival curve of the multitarget model can be characterized by
   a. D_1
   b. D_0
   c. n
   d. All of the above
3. The multitarget–single hit model can be characterized by
   a. D_1
   b. D_0
   c. n
   d. All of the above
4. Target theory was developed by
   a. Lea
   b. Orton
   c. Strandqvist
   d. Elli
5. The isoeffective curve of the time–dose model can be expressed as
   a. $D = [T/(T+G)]^{0.17}$
   b. $D = kT^{1-p}$
   c. $P(n) = [(e^{-x})(X^n)]/n!$
   d. $N = N_0^{e-D/D_0}$
6. The dose required to reduce the cell surviving fraction from 0.1 to 0.037 is referred to as
   a. D_1
   b. D_0
   c. n
   d. All of the above
7. The linear quadratic model is
   a. Empirical model
   b. Survival-based model
   c. Both
   d. None
8. Method based on the ability of surviving cells to reproduce in the culture medium after irradiation is referred to as
   a. In vitro method
   b. In vivo method
   c. Exogenous colony assay
   d. Endogenous colony assay
9. What does a "cell survival curve" describe?
   a. The relationship between the radiation dose and the number of cells that have gone through one mitosis after irradiation

   b. The relationship between the radiation dose and the proportion of cells that remain clonogenic
   c. The relationship between the radiation dose and the number of cells that have not suffered the loss of a specific function
   d. The relationship between the radiation dose and the proportion of cells that can produce DNA

10. The survival curve corresponding to a single target–single hit model is
    a. Linear
    b. Exponential
    c. Linear quadratic
    d. None

9

Biological Basis of Radiotherapy

Objective

The objective of this chapter is to get students familiar with (1) radiobiological issues introduced in modern radiotherapy techniques and (2) the application and use of clinical radiobiology in radiotherapy treatment planning. Concepts such as biologically effective dose (BED), equivalent uniform dose (EUD), and routine example calculations are described. Current and future status of radiobiological modeling in plan optimization is also discussed in order to get familiar with the advantages and limitations of such mathematical models.

9.1 Radiobiological Aspects of Modern Radiotherapy Techniques

Modern external-beam radiotherapy is usually given by linear accelerators producing high energy X-ray beams of 4–25 megavoltage (MV). There are other classes of radiation that are being increasingly adopted in radiotherapy:

- Light particles (protons and α-particles)
- Heavy particles (e.g., carbon ions)
- Neutrons

Apart from very different depth–dose absorption profiles compared with uncharged particles (i.e., neutrons) or conventional X- and γ-rays, light and heavy particles may have a greater biological effect per unit dose than conventional X- and γ-rays do. Charged particle beams are associated with a higher linear energy transfer (LET) than MV X-ray beams. High-LET radiations have an increased relative biological effectiveness (RBE), which is the highest at a very small dose per fraction and decreases with increasing fractional dose.

Among modern radiotherapy techniques, brachytherapy plays an important role due to various overall treatment times and short treatment distances. Table 9.1 summarizes all alternative radiation techniques to date.

TABLE 9.1

Radiation Techniques Implemented to Date

Radiation Therapy	Particle	Implementation
Brachytherapy	Isotope based	Ultra low dose rate, LDR, HDR, Pulsed brachytherapy
Particle	Proton/Heavy ion	Sychnotron/heavy ion centers
Cobalt-60 based	Photon	SRS-Gamma Knife®
Linear accelerator	Photon	3D-CRT, IMRT/IGRT, SRT Tomotherapy® Cyberknife® X-Knife®

9.1.1 Brachytherapy

Brachytherapy may be delivered at high dose rate (HDR), medium, or low dose rate (LDR) or in the form of pulsed brachytherapy. Permanent seed implants represent a special case of LDR brachytherapy during which the dose rate falls continuously. Potential radiobiological advantages of brachytherapy include varying treatment times. Short treatment time could prevent tumor repopulation while longer treatment time could redistribute cells into sensitive cell cycle phases and also allow reoxygenation with time after implantation.

In brachytherapy, the main radiobiological considerations are related to dose rate and fractionation effects and to the correspondence between the two. Radiobiological quantification of brachytherapy applications may be performed with extended versions of the linear quadratic (LQ) model developed from the same biophysical bases as the more familiar external beam model (Dale, 1985). However, the rapid dose fall-off around individual sources (or source dwell positions) creates a more complex interplay between the physical and radiobiological aspects than in the case of external beam treatment. This means that the biological effects associated with brachytherapy depend largely on the source geometry and on the exact location of the dose prescription points. These aspects should be borne in mind when comparing different treatments (Dale et al., 1997; Dale and Jones, 1998; Armpilia et al., 2006).

An important parameter included in quantification of LDR brachytherapy is the repair rate (μ value), which refers to the repair process of DNA sublethal damage (SLD). There is some evidence that tumor repair half-lives may be shorter than those for late-responding tissues. If the repair process is assumed to be exponential in form, then the repair rate (μ) is related to the repair half-time ($T_{1/2}$) by the following equation:

$$\mu = \frac{0.693}{T_{1/2}}$$

Most brachytherapy dose distributions (especially in gynecological cases) are strongly influenced by the high dose gradients which are present with implications for the associated radiobiology. Methods exist (Dale et al., 1997; Armpilia et al., 2006) for incorporating dose gradient effects. Due to high dose gradients, the anatomical location of dose specification points thus remains fundamentally important. With the advent of image-guided brachytherapy we have moved from point-based to volume-based dose prescribing. This has resulted in some differences in the application of radiobiology to clinical practice with an increased importance of the understanding of the role of brachytherapy prescribing.

9.1.1.1 Brachytherapy Models

Biological effects and their quantification are dependent on the radiobiological characteristics of the irradiated tissues: α/β values, repair rates (μ values), and repopulation rates. Radiobiological quantification of brachytherapy applications may be performed with extended versions of the LQ model. The differential behavior of the dose rate effect between tissues is similar to the differential behavior observed in the fractionation effect. In both cases, the behavior is strongly influenced by the α/β ratio.

9.1.1.1.1 Fractionated HDR Brachytherapy

Fractionated HDR brachytherapy typically involves ~2–6 fractions delivered ~24 hours (or more) apart.

Individual dose fractions create lethal damage (fixed) and SLD (repairable). The SLD is repaired between one fraction and the next.

The BED is calculated as follows:

$$\mathrm{BED} = \mathrm{Nd} \times \left[1 + \frac{\mathrm{d}}{(\alpha/\beta)}\right] - \mathrm{k}\left(\mathrm{T} - \mathrm{T}_{\mathrm{delay}}\right)$$

where

N is the number of fractions

d is the dose per fraction

This equation is valid only if fractions are sufficiently well spaced (usually a minimum of 6 hours) to allow complete recovery of SLD between successive fractions.

9.1.1.1.2 Continuous LDR Brachytherapy

Continuous LDR (CLDR) brachytherapy radiation delivery takes hours/days, and therefore some of the SLD gets repaired while the radiation is being delivered. Moreover, since LDR brachytherapy treatments last up to a few days, the repopulation effect is usually negligible. The BED equation for a total dose D may be written as

$$\mathrm{BED} = \mathrm{D}\left[1 + \frac{\mathrm{Df(t)}}{(\alpha/\beta)}\right]$$

where f(t) is the function which quantifies the reduction in lethality as treatment time is increased.

For mono-exponential repair (μ is the time constant) the BED equation may be written (Dale, 1985; Thames, 1985) as

$$\mathrm{BED} = \mathrm{RT} \times \left[1 + \frac{2\mathrm{R}}{\mu(\alpha/\beta)}\left(1 - \frac{1 - e^{-\mu \mathrm{T}}}{\mu \mathrm{T}}\right)\right]$$

where

R is the dose rate (Gy/hour)

T (hour) is the treatment time

μ (hour^{-1}) is the time constant for mono-exponential repair (Dale, 1985; Thames, 1985)

9.1.1.1.3 Permanent Implants

In the case of permanent implants, the physical decay of the radioactive sources directly affects the rate of production of both lethal and sublethal damage and therefore has a direct impact on the relative effectiveness of such treatments. In the absence of any repopulation effects, it can be shown (Dale, 1985) that

$$BED = \frac{R_0}{\lambda}\left[1 + \frac{R_0}{(\mu+\lambda)(\alpha/\beta)}\right]$$

where

R_0 is the initial dose rate
λ is the radioactive decay constant

In treatments involving long-lived radionuclides, radiation is delivered over a long time period. Thus, even for slowly growing tumors, there may be appreciable repopulation, which will effectively reduce the net BED, and should therefore be taken into account. Additionally, because the dose rate continually falls during treatment, there will come a point at which the dose rate exactly matches that the value required to sterilize any remaining tumor clonogens. The likelihood of continuing to kill cells therefore reduces substantially after this time, which is referred to as the effective treatment time (T_E) of the implant (Dale, 1989).

9.1.1.1.4 Dose Gradient Effects on Biological Response

In most brachytherapy applications, the dose prescription point will be part of an isodose surface containing volumes of tissue where the dose rises rapidly in the vicinity of the sources (or source dwell positions). BEDs specified at a single point do not account for the dose gradient effect and should therefore be used with caution, particularly when comparing treatments with different source geometry. Methods exist for incorporating the dose gradient effect (Dale et al., 1997; Armpilia et al., 2006) in order to determine the BED, an analogous concept to EUD for external beam therapy.

9.1.2 IMRT/SRT/IORT

Radiobiological issues involved in new treatment modalities such as intensity modulated radiation therapy (IMRT)/stereotactic radiotherapy (SRT)/intraoperative radiotherapy (IORT), which involve high doses delivered in a few or even a single fraction, mainly concern the effects of the following:

- Nonhomogeneous dose distributions.
- Fraction duration: Long fraction delivery times (>15 minutes) could be less effective on tumors than faster deliveries. This is because tumor SLD repair times may be quite short (a few minutes) and repair takes place within the fraction, thereby lessening the biological effect (Moiseenko et al., 2007). RapidArc, volumetric modulated arc therapy (VMAT), and tomotherapy may offset this effect.
- High dose per fraction: This potentially eliminates repopulation of residual tumor cells that may occur during wound healing before postoperative radiotherapy can begin. However, the therapeutic window between tumor control and adverse

reaction in normal tissue clinically established in fractionated radiotherapy may be decreased because of the higher sensitivity to changes in the fraction size for late normal tissue reaction.

- Variable radiosensitivity: Tumors are heterogeneous in many of their biological properties, including their sensitivity to radiation. The presence of such inter- and intra-tumor heterogeneity can have significant consequences for tumor curability.

These treatment modalities involve the following radiobiological issues (Mundt and Roeske, 2005):

1. The increased involvement of quantitative clinical radiobiology with the prescription and treatment planning process
2. The increased need for reliable information on dose–volume (DV) effects, to safely guide prescriptions
3. The increased need for information about the usefulness of nonuniform target dose distributions

9.1.2.1 Radiobiology Considerations of Hypofractionation (e.g., SRS and SBRT)

The topic of hypofractionation for nonpalliative therapy includes treatment modalities such as IORT, stereotactic radiosurgery (SRS), and stereotactic body radiation therapy (SBRT). More moderate hypofractionated regimens are associated with breast and prostate cancer. On the one hand, technological advances allow for the discreet placement of high doses of radiotherapy with limited involvement of normal tissue. Such large-dose hypofractionated treatments, however, have raised a number of significant biological questions. Underlying mechanisms for tumor killing with high dose (>8 Gy) per fraction are still not well understood. The most popular point/counterpoint discussion relates to the common use of mathematical models to describe both biological and clinical aspects of radiation response. Many believe that cell-survival curves begin to straighten at very high doses so cell survival will be greater than that predicted by the LQ model. Others believe that vascular damage after high doses increases the effectiveness of radiation and hence cell survival will be less than that predicted by the LQ model. The use of α/β ratios at such high dose per fraction modalities has also been questioned. Modest-dose SBRT experimental evidence tends to indicate that the LQ model is sufficient. Another important factor when comparing SBRT with conventional radiotherapy seems to be reoxygenation since large fraction size SBRT may prohibit reoxygenation of hypoxic tumor cells.

Clinical trials (which started with SRS of the brain but also for other sites) have shown that with highly conformal therapy, hypofractionation can be at least as effective as conventional fractionation both for cure and avoidance of normal tissue complications. However, more clinical trials are essential in order to study the efficacy of single doses/few fractions in radiation therapy.

9.1.3 Protons, High-LET Sources, and Boron Neutron Capture Therapy

Proton therapy has been administered for various tumor entities (chordoma, chordosarcoma, glioblastoma, prostate) and malignancies (head, neck, gastrointestinal, and lung tumors) which require precise, highly conformal dose deposition (Brada, Mc Donald). Heavy ion therapy is more experimental than proton therapy. Proton or heavy ion radiotherapy may be preferred for the treatment of childhood malignancies because of the reduced normal tissue exposure and thus the reduced risk for the induction of secondary cancers. Doses for ion radiation are specified in GyE, which is the absorbed dose multiplied by the RBE value.

High-LET forms of therapy using charged particle beams offer superior physical dose distributions and are also associated with a higher RBE than X-rays. The highest RBEs are associated with heavy ions rather than protons (RBE = 1.1.) and those beam types have the added advantage of a reduced oxygen enhancement ratio, potentially making them suitable for treatment of radio-resistive, hypoxic tumors. For high-LET radiations, the RBE is the highest at very small dose per fraction and decreases with increasing fractional dose. Radiobiological modeling incorporating RBE effects in the LQ model have been developed and discussed elsewhere (Dale 1989, Dale and Jones 2007, Carabe 2007) and also in Chapter 8.

9.1.3.1 Boron Neutron Capture Therapy

The radiation doses delivered to tumor and normal tissues during boron neutron capture therapy (BNCT) are due to energy deposition from three types of direct ionizing radiation that differ in their LET, which is the rate of energy loss along the path of an ionizing particle:

1. Low-LET γ-rays, resulting primarily from the capture of thermal neutrons by normal tissue hydrogen atoms [$^{1}H(n,\gamma)^{2}H$]
2. High-LET protons, produced by the scattering of fast neutrons and from the capture of thermal neutrons by nitrogen atoms [$^{14}N(n,p)^{14}C$]
3. High-LET, heavier charged alpha particles (stripped down ^{4}He nuclei) and lithium-7 ions, released as products of the thermal neutron capture and fission reactions with ^{10}B [$^{10}B(n,\alpha)^{7}Li$]

Since both tumor and surrounding normal tissues are present in the radiation field, even with an ideal epithermal neutron beam, there will be an unavoidable, nonspecific background dose, consisting of both high- and low-LET radiation. However, a higher concentration of ^{10}B in the tumor will result in it receiving a higher total dose than that of adjacent normal tissues, which is the basis for the therapeutic gain in BNCT. The total radiation dose (Gy) delivered to any tissue can be expressed in photon-equivalent units as the sum of each of the high-LET dose components multiplied by weighting factors (Gy_w), which depend on the increased RBE of each of these components.

9.2 Biological Treatment Planning

9.2.1 Tumor Control Probability and Normal-Tissue Complication Probability Curves

The probability of local tumor control by radiotherapy increases with dose according to a sigmoid relationship (Brahme, 1984) (Figure 9.1). The probability of normal-tissue damage also increases with dose. The region between the curves is called a therapeutic window and describes the difference between the tumor control dose and the tolerance dose. The farther the normal-tissue complication probability (NTCP) curve is to the right of the tumor control probability (TCP) curve, the larger the therapeutic index, that is, the tumor response for a fixed level of normal tissue damage.

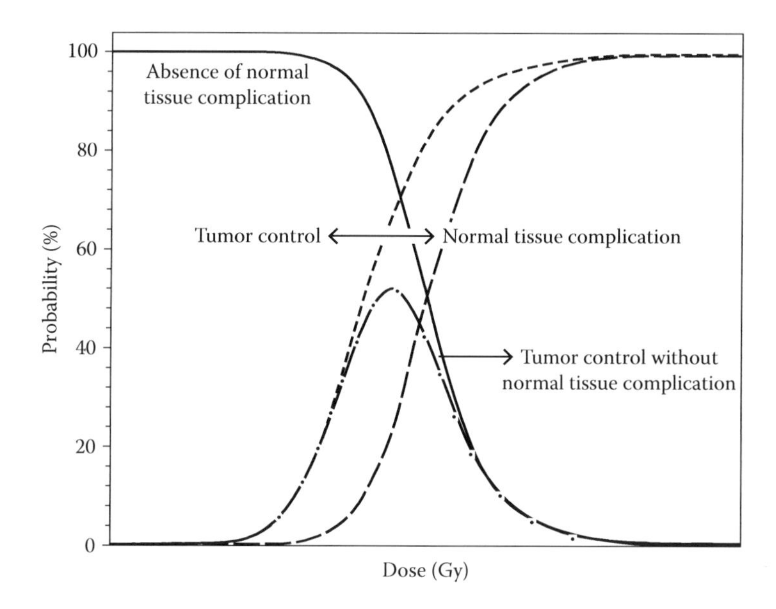

FIGURE 9.1
Dependence of TCP and NTCP on dose. The probability of tumor control without normal tissue complications receives its maximum in the so-called therapeutic window. (With kind permission from Springer Science+Business Media: *New Technologies in Radiation Oncology*, 2006, 221, Schlegel, W., Bortfeld, T., and Grosu, A.-L., eds., Figure 18.1.)

The steepness of dose–response curves is small when dose (and effect) is very low or very high and it reaches a maximum round about 50%. The theory and significance of dose–response curves has been well described by Bentzen and Tucker (1997).

The aim of radiotherapy is to give sufficient dose to the tumor to achieve local control with minimal (damage) complications to the surrounding normal tissue. The optimum choice of radiation dose delivery is such that it maximizes the TCP and simultaneously minimizes the NTCP. For a typical radiotherapy treatment, $TCP \geq 0.5$ and $NTCP \leq 0.05$.

9.2.1.1 Factors Affecting Slope of TCP Curve

In an ideal model situation where there is no variability between tumors or patients, tumor control would be achieved when the last clonogenic cell is sterilized. The probability of cure is therefore based on a Poisson distribution. If Ns is the number of clonogenic tumor cells that survive treatment, then the TCP is given by the zero-order term of a Poisson: $TCP = e^{-Ns}$.

The conventional way of expressing the steepness of a dose–response curve is to indicate, at the steepest point of the curve, the percentage increase in effect for 1% increase in dose, the so-called γ-value (Brahme, 1988). For example, the steepness is γ50 at the 50% effect level and γ37 at the 37% level. The maximum γ-value for a tumor whose response is purely determined by Poisson statistics depends on the total number of clonogenic cells ($\sim 10^9$–10^{12}) but is probably in the region of 7.

However, dose–response curves are less steep than the ultimate Poisson. Tumor-related factors include variations in clonogen content within a given stage of a specific site and histology of the tumor, variations in intrinsic tumor clonogen sensitivity, variability in radiosensitivity related to heterogeneity of oxygenation, and variation in the rates of growth of tumors during radiation therapy, which causes variability in the "effective" number of clonogens during sterilization. Treatment-related factors include heterogeneity of dose distribution, implementation of the treatment plan in overall treatment duration. γ-values for clinical local tumor control are often in the range of 1–3 (Brahme, 1988; Bentzen and Tucker, 1997).

The response of normal tissue to radiation dose has all of the complexity of tumor-tissue response in addition to the importance of functional relationship among cells. As discussed in Chapter 4, serial-like tissues such as the spinal cord can lose their function if a small circumferential region of the tissue is damaged. Parallel-like tissues such as the lung or liver have significant capacity so that a sizable amount of damage can be done without a complication occurring. Therefore, NTCP has an organ-specific dependence on the volume of damage.

Curves for normal-tissue complications are less well defined than tumor control curves because the treatment philosophy is to avoid a high incidence of damage. However, it is likely they are steeper than for tumor control, reflecting less heterogeneity in the biology of normal tissues than that of tumors. Thus, in practice, the curve for tumor control is always likely to be shallower, and may actually cross that of complications (as shown in Figure 9.1). The value of γ depends, also, on whether the fraction number of the dose per fraction is kept constant as the dose is increased. Dose–response curves tend to be steeper for late reactions than for early reactions.

9.2.2 Concepts of Nominal Standard Dose

As discussed in Chapter 8, the first scientific approach for correlating dose to overall treatment time to produce an equivalent biological isoeffect began with the work of Strandqvist, 1944, and Cohen, 1968. In the same direction, the Ellis, 1971 or nominal standard dose (NSD) formula (equation) was produced and clinically used for many years:

$$\text{Total Dose} = \text{NSD} \times \text{n}^{0.24} \times \text{T}^{0.11} \tag{9.1}$$

The Ellis formula is an example of an isoeffective relationship. It allows a total dose to be calculated for any given values of the number of fractions (n) and overall treatment time (T) that gives a fixed level of effect. NSD is a parameter that controls the effectiveness of the treatment. It has gradually become clear that such power-law models work well for certain well-defined conditions, in particular for early reactions to skin and within a limited range of fraction numbers. Outside this range, it seems to overestimate the tolerance dose when the number of fractions is very small or very large, a conclusion that is well supported by clinical data. Orton, 1973 and Cohen, 1968 summarized the limitations of NSD.

9.2.3 Biologically Effective Dose (BED) and Isoeffective Dose Calculations

The LQ approach to fractionation starts from a cell survival relationship of the following form:

$$\text{Survival fraction}(\text{SF}) = \exp\left(-\alpha \text{D} - \beta \text{D}^2\right) \tag{9.2}$$

where D is the total radiation dose given in n fractions (D = nd) and α and β are the linear and quadratic coefficients, respectively, which are respective measures of the initial slope and degree of downward curvature (bendiness/shape of the shoulder) of the underlying cell survival curve.

The "effect (E)" of radiation of n fractions can be considered as the log cell kill:

$$E = -\log(SF) = n\left(ad + bd^2\right) = aD + \beta dD \tag{9.3}$$

The quantity E/α (units of Gy since E is dimensionless and α has units Gy^{-1}) can be considered as the dose that would produce the effect E when given in very small dose fractions (note that in the limit $d \rightarrow 0$, $E/\alpha \rightarrow D$). This concept (originally suggested by Barendsen, 1982) constitutes the definition of BED:

$$BED = \frac{E}{\alpha} = nd\left[1 + \frac{d}{\alpha/\beta}\right], \tag{9.4}$$

or

$$BED = D\left[1 + \frac{d}{\alpha/\beta}\right] \tag{9.5}$$

Thus, **BED** is regarded as a measure of the true biological dose delivered to a tumor or organ (characterized by a specific α/β ratio) by a combination of dose per fraction and total dose. It is defined as the theoretical total dose that would be required to produce the isoeffect E using a (infinitely) very large number of (infinitely small) tiny dose fractions.

The **α/β ratio** is an inverse measure of fractionation sensitivity (i.e., sensitivity of a given tumor/organ to changes in dose/fraction) of a tissue or organ. The unit of α/β is the Gy. This term is used to describe different abilities of normal tissues and tumors to withstand fractionated radiotherapy. Late-responding normal tissues (which generally have low values of α/β = 1–5 Gy: said to have more sparing capacity) are usually more sensitive to the effects of changed fractionation than are tumors and early-responding tissues (which generally have higher values of α/β = 5–25 Gy).

Large fractionation studies in the laboratory in the 1980s produced a number of α/β ratio estimates for various normal-tissue end points. In parallel with these experimental studies, a number of clinical studies have produced α/β estimates for human end points (Fowler, 1989; Joiner, 2009) but precise values are rarely known in human individuals and ranges of α/β ratios are recommended to be used in BED calculations.

BED values are expressed in units of Gray with a suffix that denotes the numerical value of the α/β ratio used in the calculation. The bracketed term in Equation 9.5 is called the relative effectiveness (RE) per unit dose so that

$$BED = \text{Total physical dose} \times RE$$

This means that two adjacent tissues with different α/β ratios, each receiving the same dose and fractionation, will be associated with different values of BED.

BED can be used for comparing the biological effects of various radiotherapy schedules and for different types of radiotherapy. For example, if the biological effect of a total dose

D_1 derived in n_1 fractions of dose d_1 each is equivalent to the biological effect of a dose D_2 in n_2 fractions of dose d_2, then it follows that the BEDs will be equal and

$$D_1\left(1+\frac{d_1}{\alpha/\beta}\right)=D_2\left(1+\frac{d_2}{\alpha/\beta)}\right)$$

or

$$\frac{D_1}{D_2}=\frac{d_2+(\alpha/\beta)}{d_1+(\alpha/\beta)}$$

A more practical alternative in clinical practice is to calculate the isoeffective dose in 2 Gy fractions (i.e., equivalent doses delivered in 2 Gy fractions as the 2 Gy/fx represents the standard fractionation regimen)

$$EQD_2=D\cdot\frac{d+(\alpha/\beta)}{2+(\alpha/\beta)}$$

When using BED calculations to estimate the biological dose to tumors rather than to normal tissues, the following main differences are observed: (1) the numerical range of α/β ratios is wider in tumors and data are lacking for many specific tumor types; and (2) a repopulation correction factor should be included in case of tumors that contain rapidly proliferating clonogens.

To allow for the effect of tumor proliferation, the BED is calculated as follows:

$$BED=nd\left[1+\frac{d}{\alpha/\beta}\right]-K\cdot(T-T_{delay})$$

where

T is the overall treatment time

T_{delay} is the delay time (from the beginning of treatment) before the onset of significant repopulation

K (Gy/day) is the biological dose per day required to compensate for ongoing tumor repopulation

Tumors possessing a high K factor repopulate rapidly and treatment interruptions may be especially critical. K is related to potential doubling time T_{pot} according to the equation

$$K=\frac{\ln 2}{\alpha\cdot T_{pot}}$$

Typical values of K used for squamous cell cancers are 0.5–0.9 Gy/day (Jones et al., 2001). The conventional range of T_{delay} based on accelerated repopulation clinical data sets is between 21 and 28 days (Jones et al., 2001).

A modified BED formulation may be required in cases where inter-fraction intervals are short, as occur in accelerated fractionation. If there is incomplete SLD repair between fractions, this leads to an enhanced BED value given by

$$\text{BED} = \text{nd}\left[1 + \frac{(1+\text{h})\cdot \text{d}}{(\alpha/\beta)}\right]$$

where h is the incomplete repair correction factor. Values of h have been computed for mono-exponential repair kinetics for equally spaced fractions at various repair rates (Thames, 1985; Joiner, 2009).

9.2.3.1 General Considerations in the Use of BED

- A single one-point BED calculation will not be representative of the biological effect throughout a large planning target volume (PTV) or critical structure where there are significant hot spots.
- Hypofractionation causes increments in both dose per fraction and the total dose at the hot spot relative to the prescription point (referred to as the "double trouble" effect by Lee et al., 1995). Biological doses are enhanced proportionally more than the physical dose at hot spots (due to the quadratic term in the LQ model).
- BEDs are additive when there is a need to combine radiotherapy treatments or phases of treatments, although the volume of tissue is not inherently included in BED calculation and the high-dose region needs to be separately assessed.

9.2.3.2 Clinical Applications/Example Calculations

Example 9.1: Converting a Dose into the Isoeffective Dose in 2 Gy Fractions

A patient with bone metastasis located at the third thoracic vertebra is intended to receive 10 fractions of 3 Gy/f. What is the isoeffective dose in 2 Gy fractions for the spinal cord? From the plan dose distribution, it can be seen that there is a hot spot of 110% in the contoured spinal cord.

Considering the 110% hot spot in the spinal cord (with $\alpha/\beta = 2$), the maximum physical daily dose for the spinal cord equals 3 × 1.1 = 3.3 Gy. Therefore, the isoeffective dose in 2 Gy/f is calculated as

$$\text{EQD}_2 = \text{D}\cdot\frac{\text{d}+(\alpha/\beta)}{2+(\alpha/\beta)} = 3.3\times 10\times\left(\frac{3.3+2}{2+2}\right) = 43.73\ \text{Gy}$$

If there was no hot spot in the spinal cord, then 30 Gy delivered in 10 fractions would be biologically equivalent to $3\times 10\times\left(\frac{3+2}{2+2}\right) = 37.5\ \text{Gy}$. Thus, the hot spot of 110% in physical dose resulted in raising EQD_2 by 117%.

Example 9.2: Comparison between Different Radiotherapy Schedules

A breast cancer patient is intended to receive hypofractionated radiotherapy with 53.4 Gy in 20 fractions. Compare the hypo schedule with the conventional one of total dose of 60 Gy in 30 fractions for tumor control, early and late effects.

- The α/β ratio for breast cancer is considered to be low, at 4–5 Gy.
- For late effects (e.g., fibrosis), α/β is considered to be 3 Gy whereas for early effects (e.g., skin erythema, dry desquamation), α/β is considered as 8 Gy.

The BED for tumor was calculated according to the formula derived from the LQ model including the repopulation effect correction. This correction must be taken into account for postoperative breast tumors treated with radiotherapy because surgical resection can leave behind just a handful of viable cells, which, because they are then well vascularized, are capable of rapid growth. This is confirmed in the clinical data and discussed elsewhere in the literature (Wyatt et al., 2008; Jones et al., 2001).

$$BED = nd\left(1 + \frac{d}{\alpha/\beta}\right) - K(T - T_d)$$

where
d is the fraction size (Gy)
n is the number of fractions

- $\alpha/\beta = 4$ Gy
- T (overall treatment time) = 40 days (6 weeks) for the conventional scheme of 30 fx
  = 26 days (4 weeks) for the scheme of 20 fx
- T_d (delay time to onset of accelerated repopulation). From the literature (Wyatt), an effective doubling time T_{eff} of 26 days is assumed to start immediately after surgery and T_d is considered as zero (Ramsey et al., 2010; Sanpaolo et al., 2012).

K (Gy/day) is the biological dose per day required to compensate for ongoing tumor cell repopulation, calculated based on T_{pot} (potential doubling time) and α (radiosensitivity coefficient): $K = \ln 2/(a \cdot T_{pot})$. From the literature (Qi et al., 2011; Haustermans et al., 1998; Wyatt et al., 2003), $T_{pot} = 14$ days and $\alpha = 0.08$.

Table 9.2 provides the BED calculation for both the hypofractionation scheme and the standard fractionation scheme. BED calculation for breast tumor given in the first column includes the appropriate correction for repopulation.

By comparing the two schedules one can notice that the hypofractionation schedule results in 11% greater BED value for tumor and 5% lesser for acute effects than for the conventional scheme. Both schemes show similar BED results for late effects.

This example highlights the advantage of hypofractionation when tumor α/β ratio is low (i.e., <5 Gy).

TABLE 9.2

Resulting BED Values for Two Different Radiotherapy Fractionation Schedules in Breast Cancer

Treatment Schedule	BED (Gy) Tumor (α/β = 4 Gy)	BED (Gy) Late Fibrosis (α/β = 3 Gy)	BED (Gy) Erythema (α/β = 8 Gy)
A (30 fx, 2 Gy/fx)	66.0	100	75
B (20 fx, 2.67/fx)	73.4	100.9	71.2

9.2.4 Treatment Gap Correction

A problem frequently addressed in clinical practice is the management of unscheduled radiation treatment interruptions, caused either by patient-related factors or by technical factors (e.g., equipment breakdown). Radiobiological and clinical evidence has shown that prolongation of overall treatment time can produce losses in local tumor control because of tumor cell repopulation. This evidence is the strongest for squamous cell carcinoma (SCC) of the head and neck, non-small-cell lung cancer, and cancer of the uterine cervix (1%–2% loss of local control per day of treatment extension). Treatment schedules may be adjusted by accelerating radiotherapy after the gap. One method of treatment compensation is to add the missed fractions to the remainder of the course either by treating on weekends or by giving multiple fractions per day separated by the maximum practical interval (at least 6 hours). Hyperfractionated compensation schemes may require a dose reduction if the probability of late reactions is kept fixed when repair is slow for normal tissues. Alternatively, delivering the remaining part of the treatment with hypofractionation may be considered. Whether the method of choice will lead to increased late effects or decreased tumor control depends on the α/β values for the related normal tissues and tumor.

Interrupted treatments need to be evaluated on an individual basis and there is no universal method for tackling all problems. Once an unscheduled gap has occurred, the first step is to determine the remaining treatment time and the number of fractions still to be delivered. If there are ways of delivering these fractions that will allow the originally prescribed treatment to be maintained (e.g., by treating on weekends or by giving all or part of the remaining treatment twice daily), then further radiobiological calculations may not be necessary. If this option is not feasible, there are a series of general steps, which are prescribed analytically elsewhere in the literature (Dale et al., 2002; Joiner, 2009), to follow based on the calculation of BED for the tumor and the normal tissue for both the prescribed schedule and the modified one chosen from the various treatment options (e.g., twice-daily fractionation, increased fraction sizes, etc.) which will be likely to produce the minimum extension to the treatment time.

Example 9.3: Gap Correction

A patient with bladder cancer is to receive 66 Gy/33 fractions/45 days. Due to machine breakdown no treatment was given during the sixth week on the 29th and 30th fractions. In order to finish as planned in 45 days, increasing the dose of the last three fractions is considered. What is the required dose per fraction for the last three fractions? What is the accompanying change in the risk of rectal complications from this modified treatment schedule? Assume $\alpha/\beta = 10$ Gy for bladder and $\alpha/\beta = 3.5$ Gy for rectal complications.

First, we calculate the tumor and normal tissue BED for the prescribed schedule. For bladder cancer a certain rate of repopulation (dose increment of 0.36 Gy/day) is required to compensate for repopulation after a lag period of 4–6 weeks. Thus,

$$BED_{tumor} = D_{total}\left(1+\frac{d}{(\alpha/\beta)}\right) - K(T - T_{delay}) = 66\cdot\left(1+\frac{2}{10}\right) - 0.36\cdot(45-20) = 70.2\ \text{Gy}$$

$$BED_{rectal} = D_{total}\left(1+\frac{d}{(\alpha/\beta)}\right) = 66\cdot\left(1+\frac{2}{3.5}\right) = 103.7\ \text{Gy}$$

In order to maintain tumor BED of 70.2 Gy, for the whole schedule delivered,

$$\text{BED}(\text{pre}-\text{gap}) + \text{BED}(\text{post}-\text{gap}) - \text{Tumor repopulation factor}$$
$$= \text{Required planned BED}$$

$$BED_{tumor} = 28\cdot 2\cdot\left(1+\frac{2}{10}\right) + 3\cdot d\left(1+\frac{d}{10}\right) - 0.36\cdot(45-20) = 70.2\ \text{Gy}$$

where d is the new value of dose per fraction to be utilized over the last three fractions. The solution for d in the given equation is d = 3.06 Gy, that is, the last three fractions should be increased from 2 to 3.06 Gy in order to restore the tumor BED to that the value initially prescribed.

For normal tissue, the compensated treatment increases BED_{normal} to

$$BED_{rectal} = 28\cdot 2\cdot\left(1+\frac{2}{3.5}\right) + 3\cdot 3.06\left(1+\frac{3.06}{3.5}\right) = 88 + 17.2 = 105.2\ \text{Gy}$$

Thus, the revised treatment delivers an excess of 1.5% to normal tissue, which is clinically acceptable. In other cases, if it is found to be too high, it might be possible to split the difference, that is, to aim to achieve a tumor BED which is a little less than that prescribed while accepting a small increase in normal tissue BED. Such a result may be arrived at by a process of trial and error of different values of dose per fraction.

9.2.5 Concepts of Effective Uniform Dose (EUD)

The LQ model from which the BED concept is derived does not intrinsically include for allowance of the volume effect. The methods for assessing volume effects are more complex and use integrated BED and BED DV histograms (DVHs). The EUD concept was proposed by Niemerko (1997) and is defined as the uniform dose (Gy) for any dose distribution, which when distributed uniformly across the target volume would cause an equivalent survival fraction quantity. Its basis is the power-law behavior of tissue response with dose, and it has one parameter to be fitted depending on the tissue and irradiation characteristics. To extend the concept of EUD to normal tissues, Niemerko (1999) proposed a phenomenological formula referred to as the generalized EUD, or gEUD:

$$gEUD = \left(\sum_{i=1} v_i D_i^{\alpha}\right)^{1/\alpha}$$

where v_i is the partial volume with absorbed dose D_i. The free parameter, α (tissue-specific), is fitted according to published clinical data in which it is positive for normal tissue (α approaches unity for parallel normal tissues and positive infinity for serial normal tissues) and negative for tumors.

The equivalent uniform biological effective dose (EUBED) was proposed by Jones and Hoban (2000) to allow variations in the dose per fraction within a dose distribution. The EUDBED is then the BED giving an equivalent survival fraction to a given heterogeneous BED distribution:

$$\mathrm{EUDBED} = -\frac{1}{\alpha}\ln\left(\sum_{i=1}^{N} v_i e^{-\alpha \mathrm{BED}_i}\right)$$

where BED_i is calculated for N total number of voxels (bins in the differential DVH) of fractional subvolume v_i.

9.2.6 Limitations of Dose–Volume-Based Treatment Planning

With the implementation of three-dimensional conformal radiotherapy (3D-CRT), DVHs have proven useful as a tool for the evaluation and comparison of treatment plans. DV-based criteria in terms of limits on the volumes of each structure that may be allowed to receive a certain dose or higher are clinically the most widely used in a routine treatment-planning process. However, despite ample clinical data for some organs, it remains unclear which of the DVH-derived parameters is optimal for the prediction of NTCP (Rodrigues et al., 2004). Typically, values that are higher than the DV metric (e.g., V5, V20, and mean dose for lung) correlate with complication incidence. However, with typically one to three constraints, a range of optimized normal tissue DVHs comply well with these few constraints, but may still produce a different risk of complications.

Moreover, the loss of information on the spatial distribution in a DVH is a serious constraint in determining the relationship between local tissue damage and overall morbidity. A high-dose region in the histogram may represent a single hot spot in the volume of interest or a number of smaller hot spots from contiguous regions or from different regions. These could have considerably different implications for tissue tolerance. Specifying multiple DV constraints increases complexity of the inverse treatment-planning procedure. In IMRT planning, as plan complexity increases, dose heterogeneity becomes significant. Thus, spatial dose distribution inside a structure becomes important, especially when cold spots (underdosed regions) or hot spots (overdosed regions) are present.

DVHs also do not differentiate between functionally or anatomically different subregions or compartments within an organ. This becomes particularly relevant if variations in radiosensitivity and/or functional consequences within the organ are evident, such as in lung, kidney, parotid gland, or particularly in the eye and the brain.

Another point is that physical dose distribution may not be the best representation of plan quality since it does not contain any of the biological information which is important in determining the treatment outcome.

To overcome the limitations of DV-based criteria, they may be supplemented with biological (or dose–response-based) criteria. Incorporating biological information in addition to physical information into the radiotherapy treatment planning is important for plan evaluation comparison and optimization.

9.2.7 Uses of Biological Models in Treatment Planning

The case for radiobiological optimization of treatment plans starts with the fact that in modern radiotherapy techniques it may not be sufficient to specify the objectives of optimization purely in terms of the desired pattern of the dose. The objectives must also include DV effects and biological indices. The basic idea in radiobiological optimization of treatment plans is shown in Figure 9.1. The aim is to achieve the maximum probability of free local control. Biological models can be grouped on empirical or theoretical basis. The empirical models derive from fitting dose–response curves to actual clinical data. Alternatively, theoretical models attempt to mathematically formulate the underlying biological process. Mechanistic (theoretical) models are often considered preferable, as they may be more rigorous and scientifically sound. However, the underlying biological processes for most tumor and normal tissue responses are fairly complex and are often not fully understood, and it may not be feasible to accurately and/or completely describe these phenomena mathematically. On the other hand, empirical models are advantageous since they typically are relatively simple compared to mechanistic models. Their use obviates the need to fully understand the underlying biological phenomena. Limitations of such empirical approaches are that they strive for mathematical simplicity and thus are limited in their ability to consider more complex phenomena. Further, it may be somewhat risky to extrapolate model predictions beyond the realm within which the model and parameter values were validated.

Modern developments in the high-precision delivery of radiation, particularly IMRT and tomotherapy, or in the administration of protons or heavy ions have stimulated clinical trials on dose escalation, notably in lung and prostate. Data are now available for tolerance to partial-organ irradiation at doses well above the levels that were previously established. The new data obtained from prospective dose-escalation studies, combined with precise knowledge of dose distributions and DVHs, should in the future allow the derivation of more realistic parameters and a validation of mathematical models describing volume effects.

The radiobiologically based indices mostly used in biological treatment planning are TCP, NTCP, and EUD. Each equation relating to a different end point can be referred to as an "objective function term." The EUD model has the advantage of fewer model parameters (one parameter for gEUD) as compared to TCP/NTCP models and allows more clinical flexibility. The proper calibration of a TCP/NTCP model requires monitoring the outcomes for a large number of patients. However, the utility of EUD for evaluating a single plan is limited. TCP/NTCP models can provide direct estimates of outcome probabilities, which are more clinically meaningful than EUD.

9.2.7.1 NTCP Complication Probability Models

The goal of NTCP modeling is usually to find a function and corresponding parameters which best predict the risk of a given complication.

The calculation of NTCP differs from the calculation of TCP due to the difference in the intended end points. For normal tissues, the relative volume of the organ irradiated to certain dose levels is used to predict the probability of causing complications. The DV relationship will differ for each organ and will generally be a mix of two extreme cases, where either a critical dose must be reached or a critical volume must receive a certain dose before complications arise.

Normal tissue is constructed of functional subunits (FSUs). The irradiation of a single FSU to a certain dose level might alter the functionality of the organ and cause complications. If this is the case, this organ will be classified under the parallel function (or critical element) model or as having a serial architecture. Clinically, the DV effect has been quantified for many organs (e.g., lung possesses a strong volume effect while esophagus and rectum are likely to be highly serial).

In clinical practice, normal tissue will not be uniformly irradiated as assumed by the Lyman model (1985); therefore, the model was extended by introducing histogram-reduction algorithms, which transform the initial multistep DVH (having the maximum dose D_{max}) obtained for a specific treatment plan into a biologically isoeffective single-step histogram with an effective volume (V_{eff}) at the dose D_{max}. The Lyman–Kutcher–Burman (LKB) model may be viewed as one in which the average functional damage must accumulate above a given level (TD_{50}) before a complication is likely.

For some tissues, high doses to small volumes may not lead to complication, though local function may be destroyed. This biologically based insight can be codified with the parallel-function models, which attempt to directly estimate functional damage by assuming that local functional damage follows a sigmoid curve. The ability of some tissues to receive at least some very high doses of radiation without manifesting a clinical end point may correspond to either (1) eventual healing of a highly damaged, but small, volume or (2) reduction of functional capacity, but not beyond that needed to avoid a complication, as may be the case for lung irradiation.

Thus, the essential difference between the parallel-function and LKB models is that, in the LKB (power-law) model, increasing the dose to one or more voxels will always change the response probability whereas for the parallel function model, increasing the dose to one or more voxels eventually causes a saturation effect due to the sigmoidal response function, and therefore an increase in dose to saturated/high-dose voxels will cause a little increase in response probability.

The majority of current NTCP models are DVH-based and therefore ignore important information about the location of cold and hot spots within an organ at risk. The so-called cluster models (Thames et al., 2004; Tucker et al., 2006) are based on the assumption that not only volume but also spatial distribution of hot spots affects complication risks. These models provide a first step toward a new class of NTCP models that would take into account the entire 3D dose distribution, and may further improve the accuracy of NTCP estimates.

At present, Quantitative Analysis of Normal Tissue Effects in the Clinic (QUANTEC, 2010) summarizes the current knowledge of DV dependencies of normal tissue complications from external beam radiotherapy and, where possible, gives quantitative guidance for clinical treatment planning and optimization. Where available, NTCP models have been compiled. This information is expected to provide a boost for further deployment of biological models in the clinical treatment-planning process.

9.2.8 Advantages of Biological Modeling over DV Evaluation Criteria

Radiobiologically relevant objective functions may increase the quality of the treatment plans by driving plans further in the direction of improvements compared to DV constraints fixed in an a priori fashion. Unfortunately, there is no guarantee that a biologically related model does indeed estimate the consequence of dose distributions if they deviate greatly from the baseline data set that led to the model parameters. However, for the purpose of dose optimization, it is sufficient that the use of the model can guide the optimization toward favorable dose distributions.

Another aspect of plan optimization is that the use of multiple DV criteria for plan evaluation of a single organ may become problematic as they need to be given individual priority and ideally ought to be combined into a single figure of merit to avoid ambiguities. In many cases, two or more DV points are used to evaluate a dose distribution in a particular organ. It might so happen that the dose distribution passes the evaluation test for some points and fails for others, requiring the treatment planner to prioritize different DV criteria. In contrast, biologically related models have the potential to provide an inherent prioritization of multiple DV criteria incorporated into a single figure of merit.

Of course, radiobiologically relevant objective functions can be mixed with the DV constraints to prevent large hot regions or unacceptably cold spots in the target, or small hot spots in serial-function normal structures.

Finally, some mechanistic biological models incorporate terms describing radiosensitivity as a function of dose per fraction, and can thus be used to predict outcomes of different fractionation schemes. DV criteria, on the other hand, apply to a single fraction size and if the standard fractionation scheme is changed, DV-based prescription/normal tissue constraints need to be modified based on clinical experience and isoeffect calculations.

9.2.9 Precautions for Using Biological Models in Plan Optimization

To evaluate a treatment plan, accurate biological models and parameter estimates become absolutely essential. Published data are available for many tumor sites and complication types (Allen et al., 2012), affording the user a variety of choices. However, this approach is fraught with significant risks if published parameter sets are applied injudiciously without following the same practices that were used to generate the original data. Caution should be exercised if clinical and demographic characteristics of the patient population under evaluation differ substantially from those in the original patient cohort used to derive published parameter estimates. The reason is that additional variables influencing the outcome, which were not present in the original population, may be present in the evaluated patient population. When using published parameter estimates for plan evaluation, it has to be carefully verified that they apply to the appropriate end points, organ volume definitions, and fractionation schemes.

Parameter estimates should clearly be used only with the model for which they were derived. In some cases, fits to more than one model are available for the same data set. For such situations, it has been observed that different NTCP models often provide different answers to important clinical problems (Zaider and Amols, 1998; Moiseenko et al., 2000; Muren et al., 2001). It is generally not possible to determine which model is right based on observing fits to clinical data (Moiseenko et al., 2000). To resolve this situation and to ensure further progress in the use of biological models for plan evaluation, concerted efforts to select the most practical models and to create databases of parameter estimates are urgently needed. Such sets of data (e.g., the QUANTEC initiative), supported by experts in TCP/NTCP modeling, will provide a strong basis for treatment planning system (TPS) manufacturers to include biologically based evaluation tools in their products.

Most NTCP models do not include explicit descriptions of dose-per-fraction effects in the attempt to minimize the number of parameters. If the fraction size in a plan under evaluation is very different from that in the data set used to derive parameter estimates, both sets of data should be normalized to the same dose per fraction, usually using the LQ formalism. If the dose per fraction varies considerably in the patient cohort from which parameter estimates are being derived, it is reasonable to normalize all doses to some standard fractionation scheme. Even if the prescribed fraction size is unchanged, the simultaneous use of an increasing number of beams/orientations (e.g., with multifield IMRT) reduces the dose per fraction in the exposed normal tissues away from the target, compared to what they would have seen with a "conventional" plan with a limited number of beam orientations used sequentially (e.g., AP/PA beams followed by opposed oblique beams). Whether self-derived or published parameter estimates are used, it is essential to standardize the organ volume relative to which the parameter is computed. For example, the EUD or NTCP for rectum and rectal wall will differ because the dose distributions in each volume differ. The EUD or NTCP will also depend on the delineated length of rectum or rectal wall. Much more subtle is the computation of biological indices for the spinal cord, where either a standardized length has to be segmented (e.g., including all thoracic and cervical vertebrae) or the parameter is computed relative to a normalized volume. Care should be taken that for parallel organs, whose response is correlated with the mean dose, the entire organ is included in the image set and dose calculation grid.

Biological figures of merit for target volumes require much less consideration since their utility for outcome prediction is frequently limited by uncertainties of individual tumor biology. It is important to understand which aspects of a target dose distribution influence the TCP. Using the Poisson-based model with interpatient heterogeneity, various investigators have demonstrated that even very small cold spots may considerably decrease the TCP, whereas the hot spots only affect the TCP to a great extent if the volume of the hot spot is large (Sanchez-Nieto and Nahum, 1999; Tomé and Fowler, 2000, 2002). Thus, one must consider limiting PTV inhomogeneity or at least constraining the hot spots to the gross tumor volume (GTV) or clinical target volume (CTV). This can be achieved by adding physical maximum dose cost functions to optimization criteria for target volumes.

Available sets of TCP parameter estimates are less consistent than NTCP parameters in the sense that different analyses use somewhat different assumptions when deriving model parameters: fixed number of clonogens versus fixed clonogen density, inclusion or exclusion of the time factor, and so on. Strictly speaking, it is incorrect to apply parameters derived using one set of assumptions to even a slightly modified model. This poses difficulties because users wishing to integrate TCP calculations into their plan evaluation routine need to implement not only different models that were used to analyze data for different sites but also different variations of the same basic model. Efforts similar to QUANTEC are needed to summarize TCP data and to derive common sets of parameters for one or two models, which could then be built into commercial TPSs.

9.2.10 Strategies for Effective Use of Biological Models in Plan Optimization

The best measure against the hazards of using biologically motivated cost functions is to understand their effect on the dose distribution and to know the desirable properties of the final dose distribution. The overall dose distribution derived from such an optimization should be carefully inspected; one should not rely purely on DVH metrics. Each desired goal should be reflected by a specific cost function term, which should be chosen so that it is capable of controlling this particular property of the dose distribution sufficiently. Thus, the task of setting up a biologically related optimization problem becomes, in the order of increasing importance, (1) choice of sufficient cost functions; (2) choice of right types of cost functions; (3) choice of right volume effect parameters; and (4) clear idea of what features make a dose distribution acceptable or unacceptable in the clinic. For example, an organ like the spinal cord, for which the maximum dose is considered to be of the highest priority, is ideally modeled by a gEUD that is $\gg$1. This kind of model is very sensitive to high doses while it is very insensitive to low and intermediate doses. Clearly, this kind of behavior is not sought for organs like lung, where the primary objective is to spare sufficient lung volume of intermediate doses while controlling the maximum dose is only of secondary importance. Here, a gEUD model with a smaller value or a parallel complication model is a better choice, but one has to be aware that this type of model does not control the maximum dose. In order to achieve this, it needs to be complemented with either a second gEUD model with a greater parameter value or a maximum dose constraint. Notice that, in this example, the two models represent two types of complication control with different volume dependency: one aims to control volume-related complications like pneumonitis and loss of lung function, while the other tries to manage more local complications like destruction of large blood vessels or even necrosis. According to the report by the American Association of Physicists in Medicine (AAPM), three commercially available and most commonly used TPSs that employ biologically based models are selected to demonstrate their practical use.

Objective-Type Questions

1. Dependence of TCP and NTCP on dose
   a. Exponential
   b. Sigmoid
   c. Linear
   d. Quadratic
2. Model in use for radiotherapy isoeffect calculations
   a. Nominal standard dose
   b. Time–dose fractionation
   c. Cumulative radiation effect
   d. Linear quadratic
3. What α/β ratio would one use for late effects in isoeffective dose calculation?
   a. 1
   b. 3
   c. 10
   d. 15
4. What α/β ratio would one use for isoeffective dose calculation for head and neck squamous carcinoma?
   a. 1
   b. 3
   c. 10
   d. 15

5. A patient with bone metastasis located in the third thoracic vertebra is intended to receive 10 fractions of 3 Gy/f. What is the isoeffective dose in 2 Gy fractions for the spinal cord (assume that it receives 100% of physical dose)?
   a. 30
   b. 33
   c. 38
   d. 40
6. A prostate patient (assume $(\alpha/\beta)_{tum} = 2$ Gy) is going to receive a hypofractionated radiotherapy schedule of 2.3 Gy/day. How many fractions should the patient receive in order to have the same tumor control compared with the conventional regime of 70 Gy in 35 fractions?
   a. 37
   b. 35
   c. 32
   d. 28
7. What action would you take for a treatment gap of 3 days in a radiotherapy schedule?
   a. Extend treatment by adding missing fractions
   b. Increase dose for the rest of the treatment
   c. Do nothing
   d. Use multiple fractions per day until end of treatment
8. What are the limitations of LQ model and BED calculation?
   a. Does not account for inhomogeneous dose distribution
   b. Not valid for doses over 7 Gy
   c. Refers only to point calculation
   d. All of the above
9. Limitations of dose–volume treatment planning
   a. Refer to single fraction size (standard fractionation)
   b. Do not contain biological information
   c. Loss of information on the spatial distribution
   d. All of the above
10. What are the radiobiological indices used in biological treatment planning and radiobiological optimization?
    a. EUD, gEUD
    b. TCP, NTCP
    c. DVH multiple criteria
    d. All

10

Biological Dosimetry

Objective

This chapter aims to study the various types of biomarkers, the processes involved, and a few related techniques suitable to quantify the radiation received by patients from occupational exposure and by the general public. It also creates a platform to perform further research in order to investigate the biological effects of radiation.

10.1 Introduction

Biological dosimetry (or biodosimetry) is the measurement of induced biological effects (biomarkers) of radiation in exposed individuals that can be related to the magnitude of the radiation dose received. These biomarkers can result from any of the following:

1. Advanced and complex medical equipment used for radiation treatment
2. X-ray machines used for diagnosis and treatment
3. Nuclear power plants
4. Industries that use radiation for food sterilization, material testing, and so on
5. Terrorist attack using radiological and nuclear devices

Biological dosimetry plays a vital role when the physical dosimetry information is not available or unreliable, if the personnel dosimeter is exposed inadvertently or maliciously, or when the accidental context is unclear. It does not measure the exposure in real time, but it indicates the cumulative exposure, radiation effects, and radio sensitivity. These are key information to confirm suspected cases of overexposure, and help the medical staff to predict the time course and severity of the radiation syndrome, deliver appropriate medical care, and assess the risk of long-term consequences. Biological dosimetry is always supplemented with clinical dosimetry, which is based on the early physical symptoms of prodromal syndrome and hematological changes.

Prodromal syndrome is a crude estimation of the absorbed dose that can be obtained from clinical diagnosis. The prodromal response to exposure to ionizing radiation is characterized by time of onset and degree of various gastrointestinal and neurovascular signs and symptoms such as nausea, vomiting, headache, fever, fatigue, weakness, abdominal pain, parotid pain, and erythema. For acute radiation sickness, vomiting is the most prominent indication. Onset of vomiting after 2 hours suggests a dose of 1–2 Gy. Vomiting between 1 and 2 hours suggests 2–4 Gy, between 30 minutes and 1 hour indicates 4–6 Gy, and within 30 minutes, especially if accompanied by explosive diarrhea, indicates that a dose over 6 Gy was received to the whole body. The occurrence of fever and chills within the first day post exposure is associated with a severe and life-threatening radiation dose. A person who has received a whole-body dose of more than 10 Gy will develop erythema within the first day post exposure. In this case, the erythema is restricted to the affected area. With lower whole-body doses that are still in the potentially fatal range, erythema is less frequently seen.

10.1.1 Hematological (Lymphocytes/Granulocytes) Changes

As discussed in Chapter 4, the hematopoietic cells in the bone marrow and peripheral blood are the most radiosensitive tissue in the body. The number of lymphocytes and granulocytes are decreased depending upon the total body dose received. Lymphocytes respond very rapidly and their level after 2 days predicts the severity of radiation injury (from none to lethal). Granulocytes respond more slowly, with extreme changes occurring approximately 20–30 days following exposure.

In clinical dosimetry, the sensitivity of these bioindicators is poor, their symptoms in sublethal doses are stable for a shorter duration, and hence biological dosimetry is recommended. The following are ideal characteristics of these biodosimeters:

1. Sensitivity in a wide dose range of 20 mGy to several Gy
2. Variable and reproducible response depending on the dose
3. Variable response to all low and high linear energy transfer (LET) radiation
4. Early and stable long-term response
5. Possibility to measure both acute and chronic exposure
6. Possibility to detect partial body irradiations and indicate the percentage of the exposed region to the total body
7. Noninvasive, that is, measurements taken on tissues or fluids that are easily obtainable
8. Possibility to repeat the procedure
9. Fast, simple, and automated technique
10. Possibility to measure the exposure few years later

10.2 Biomarkers

Nowadays, a broad spectrum of biomarkers with equally varied applications are available. Due to many challenges in all the biomarkers, there is a need to combine several biological end points so that a better and more complete picture can emerge about the exposure. The most commonly used biomarkers are briefly discussed here as given by Pernot et al. (2012):

1. Cytogenetic biomarkers
2. Biomarkers for nucleotide pool damage and DNA damage
3. Biomarkers for germline-inherited mutations and variants

4. Biomarkers for induced mutations
5. Biomarkers for transcriptional and translational changes
6. Biomarkers for epigenomic modifications
7. Others including biophysical markers of exposure

10.2.1 Cytogenetic Biomarkers

Cytogenetic biomarkers are indicators of multiple chromosomal aberrations such as dicentrics, translocations, premature chromosome condensation (PCC), telomere length, micronuclei, and complex chromosomal rearrangements (CCR). They are the most precise, highly specific, sensitive, and widely used method to analyze the biological effect of ionizing radiation. The various cytogenetic biomarkers are discussed here briefly.

10.2.1.1 Dicentrics

As discussed in Chapter 3, a dicentric chromosome is an abnormal chromosome with two centromeres. It is formed due to an exchange between the centromere pieces of two broken chromosomes. The formation of a dicentric is directly related to the amount of the dose, and hence it is mostly used as a biomarker in biodosimetry. It is a wise choice to investigate the recent exposure to ionizing radiation. As the natural formation of dicentrics is very scarce, it is possible to analyze the homogeneous whole-body exposure in the range of 0.1–5 Gy.

It is a suitable method to analyze the estimated dose of acute whole-body and partial-body exposures. However, it is not suitable for dose estimation in the case of nonuniformly distributed radionuclides in the body. This method is applicable to estimate internal radiation exposure of radionuclides such as tritium and cesium, which distribute uniformly in the body. Due to the turnover of peripheral blood lymphocytes, the occurrence of dicentrics is reduced with time and hence they are unstable. This limitation can be overcome by performing dicentric analysis within a few weeks of exposure. As mentioned earlier, it consists of the following steps:

1. Stimulation of isolated lymphocytes into mitosis by adding phytohemagglutine (PHA)
2. Arrest of metaphase chromosomes using colchicines
3. Preparation and fixation of slides
4. Staining the chromosomes with zimsa and scoring of dicentric chromosomes under the microscope

10.2.1.2 Translocations

These kinds of chromosomal aberration remain in the blood over a long period of time and hence they are used as biomarkers to analyze radiation exposure received over a long period of time. But the reading varies with respect to dose rate and whole-body versus partial-body exposure. As translocations may be parts of complex chromosomal rearrangements (CCRs), they can migrate (unstable); hence, it is difficult to confirm their origin in the cell. Therefore, only cells without unstable translocation are scored. Translocations are sensitive in the dose range of 0.25–4 Gy. The frequency of translocation occurrence also varies based on smoking, clastogenic agents in the diet or environment, gender, age, ethnicity, and genetic polymorphisms in genes involved in cellular defense mechanisms. The fluorescence in situ hybridization (**FISH**) technique is used for the detection of translocations.

10.2.1.3 Premature Chromosome Condensation

Under accidental conditions, it is important to know the amount of radiation exposure in a short period of time, which will help in the medical management of persons who have received excess exposure. In the case of dose estimation with dicentric, micronuclei, and translocation methods, it will take more than 48 hours to know the result whereas with PCC the dose can be estimated within a few hours. Moreover, at high doses (>6 Gy), the number of lymphocytes in the blood is reduced and the cell cycle is delayed. Therefore, it is not possible to perform conventional dicentric assay, which requires sufficient number of metaphase cells for dose estimation. In such a situation, a simple and cost-effective PCC technique is used that induces chromatin condensation before first mitosis. PCC can be induced two ways.

One method is by (1) fusing quiescent (G_0 phase) cells into mitotic cells using the Sendai virus or polyethylene glycol (PEG) as the fusing agent, and the other is by (2) chemically cycling cells using inhibitors of DNA phosphorylation. In quiescent cells, a number of excess PCC fragments (>46 chromosomes for human) are scored, which enables chromosomal aberrations to be seen immediately. In cycling cells, it is also possible to score ring chromosomes, dicentrics, and translocations if the PCC assay is combined with FISH chromosome painting or C-banding techniques. The dose response of this assay in quiescent cells (PCC fragment assay) is from 0.02 to 20 Gy but in cycling cells (PCC ring assay), it is sensitive from 1 to >20 Gy.

PCC is useful to detect exposure at low doses and high acute doses of low and high LET radiation. It can also discriminate between total- and partial-body exposures. Hence, PCC is a suitable method to analyze radioprotection problems and assess doses received after medical imaging (x-ray and nuclear medicine).

10.2.1.4 Telomere Length

Telomeres are the ends of linear chromosomes and are composed of tandem hexameric nucleotide repeats of the six-nucleotide sequence, 5′-TTAGGG-3′, bound to an array of specialized proteins, which are collectively termed shelterin. The main function of telomeres is to maintain the chromosomes in a stable condition by protecting their ends. Due to failure in duplicating DNA strands at it ends because of DNA polymerase malfunction, the telomere's length is reduced in each cell division. Generally, the telomere is maintained by the action of an enzyme telomerase using RNA segments to correct the reduced part. Some factors such as genomic instability, cancer progression, biological effect due to exposure, increased radiation sensitivity, loss of cellular viability, and age cause deficiency in telomerase, which leads to telomere length shortening.

Hence, precise, reproducible, and simple methods aiming to measure telomere length are needed. There are three common methods to measure telomere length: (1) southern blot, which is the gold standard method and requires large DNA quantities; (2) flow-FISH, which is a combination of flow cytometry and FISH, which requires intact cells for analysis; and (3) quantitative polymerase chain reaction (qPCR), which requires smaller quantities of DNA. All these methods are labor-intensive and time-consuming. The sensitivity of the telomere length biomarker, its suitability to analyze short-/long-term exposure, and the time of analysis are under development.

10.2.1.5 Micronuclei

Micronuclei (MN) originate in two ways: (1) from acentric fragments of chromosomes, which are not properly segregated into daughter cell nuclei at anaphase but instead remain in the cytoplasm after cell division, and (2) from whole chromosomes, which are unable to migrate with the rest of the chromosomes during the anaphase of cell division. The MN assay is a mutagenic test for the detection of chemicals/agents that modify chromosome structure and segregation and therefore induce MN in the cytoplasm of interphase cells.

MN can be seen as small spherical objects using any conventional DNA dye. In comparison to the dicentric method, MN assay is an easier technique to score bi-nucleated cells both manually and by using automated scanning and image analysis. Like dicentrics, MN is lost when cells continue to divide as it is formed during cell division. Hence, the MN assay is carried out by blocking the cell cycle progression of PHA-stimulated lymphocytes at the stage of cytokinesis after the first mitosis using cytochalasin B. The three popular methods suitable to analyze MN are (1) cytokinesis block MN assay, (2) MN centromere **FISH** assay for low doses, and (3) flow cytometric detection of DNA in reticulocytes. With regard to the sensitivity of detection, it is comparatively less sensitive to the dicentric method, with its sensitivity ranging from 0.25 to 4 Gy.

10.2.1.6 Complex Chromosomal Rearrangements

CCR refers to a combination of several chromosomal aberrations, such as translocations, dicentrics, ring chromosomes, or acentric fragments in a cell. CCRs are unstable since they decrease with the turnover of peripheral blood lymphocytes. The natural occurrence of CCR is very low but it is caused by high LET and heavy ion exposure. The multicolor **FISH** technique is used for the detection of CCRs. It should be performed as early as possible before the peripheral blood lymphocytes are renewed.

10.2.2 Biomarkers Related to Nucleotide Pool Damage and DNA Damage

As discussed in Chapter 3, ionizing radiation can induce a variety of DNA damage either directly or indirectly depending upon the type of radiation, dose, dose rate, and so on. As DNA damage is induced by many factors such as age, oxidative stress-related syndrome, smoking, and chemicals, it is not unique only to ionizing radiation. DNA strand breaks can be measured directly or by using surrogate end points such as the presence of gH_2AX foci or the comet assay. Even though there are many methods to measure nucleotide pool and DNA damage, only three are discussed here briefly.

10.2.2.1 DNA Single/Double Strand Breaks

Since single strand breaks (SSB) and double strand breaks (DSB) are highly characteristic of the DNA lesions formed after exposure, assays detecting their formation and persistence or the individual's ability to repair this type of damage can be used as biomarkers for exposure or individual radiation sensitivity. There are many techniques such as alkaline or neutral filter elution, alkaline unwinding, sucrose gradient centrifugation, or pulsed field gel electrophoresis available to analyze DNA damage. But most of these techniques are not sensitive enough to analyze the damage induced by a dose below 2 Gy.

The comet assay, on the other hand, is a relatively easy, quick, and automatable method to detect both in vitro and in vivo DNA damage and repair following exposure at the single cell level with sensitivity in the range of 0.1–8 Gy. It requires minimal number of cells (~10,000) or volume of whole blood (10 mL). The assay can be performed in neutral or in alkaline conditions. Although both methods can detect SSB and DSB, the alkaline assay is often used to detect SSB and the neutral assay to detect DSB. It can be performed within a few minutes to a few days after exposure.

10.2.2.2 gH_2AX

When DNA DSB occurs in cells, phosphorylation of the histone protein H_2AX (H_2A protein family) takes place at the site of the DSB. It forms and accumulates a constant number/percentage of gH_2AX foci in the cell nucleus depending upon the dose within a few minutes of damage. The maximum yield of foci is detected between 30 and 60 minutes following irradiation and then there is a decline in the number of foci. Hence, it is used as a biomarker to assess the formation of DNA damage. It is sensitive from 0.01 to 8 Gy dose. But it is limited due to rapid decrease of the signal and changes in the number of foci. There are two ways to perform the assay: (a) **flow cytometric analysis**, and (b) automated microscopic analysis.

10.2.2.3 Extracellular 8-oxo-dG

As discussed in Chapter 3, reactive oxygen species (ROS) (molecules or ions), such as superoxide, peroxide, and hydroxyl radicals, released continually from mitochondria during normal cellular metabolism get enhanced due to the interaction of indirectly ionizing radiation with oxygen molecules, which leads to oxidative stress. Under oxidative stress, 8-hydroxy-2′-deoxyguanosine (8-oxodG) is formed in DNA in two ways: (1) by the direct oxidation of DNA bases and (2) by getting added into DNA by DNA polymerase as a modified base drawn from the nucleotide pool. The yield of 8-oxodG serves as a biomarker for oxidative DNA damage. It is sensitive in the dose range of 0.01–8 Gy and is analyzed by high-performance liquid chromatography coupled to electrochemical detection–modified enzyme-linked immunosorbent assay (HPLC-ECD coupled with **ELISA**).

10.2.3 Biomarkers Related to Germline-Inherited Mutations/Variants

10.2.3.1 Single Nucleotide Polymorphisms and Inherited Gene Mutations

Single nucleotide polymorphism (SNP, pronounced "snip") is a DNA sequence variation that occurs commonly within a population. SNPs are single nucleotide substitutions of one base for another. Each SNP location in the genome can have up to four versions: one for each nucleotide A, C, G, and T. Not all single nucleotide changes are referred to as SNPs. If two or more versions of a sequence present in at least 1% of the general population, it is classified as a SNP.

SNPs are divided into two main categories: (1) linked SNPs (also called indicative SNPs), and (2) causative SNPs. Linked SNPs do not reside within genes and do not affect protein function. But they create the risk of getting a certain disease. Causative SNPs affect the way a protein functions, correlating with a disease or influencing a person's response to medication. Causative SNPs come in two forms: coding SNPs (located in the coding region that change the amino acid sequence of the gene's protein product), and noncoding SNPs (located within the gene's regulatory sequences that change the timing, location, or level of gene expression). Genome-wide association study (GWAS) is used to compare arrays of SNPs from cases and controls in an attempt to identify functional DNA sequence variants that influence radiation sensitivity.

10.2.3.2 Copy Number Variant and Alteration

Both copy number variant (CNV) and alternation (CNA) are a form of structural variation in the genome. These variations may be due to two reasons: (1) deletion of a large number of genomes (larger than the normal number), or (2) duplication of a large number of genomes. For example, if the normal genome is A-B-C-D-E, then it may be A-B-D-E (deletion) or A-B-C-D-D-E (duplication). CNAs can be measured with array comparative genomic hybridization (aCGH), FISH, targeted genome sequencing, and so on, if one knows the important molecular features of human genetics. Even though many studies have found that CNV can be a biomarker for analyzing radiobiological effect, it is still in the research stage.

10.2.4 Biomarkers Related to Induced Mutations

Two somatic mutation assays—glycophorin A (GYPA) and hypoxanthine-guanine-phosphoribosyl transferase (HPRTase)—are frequently used in biodosimetry. There are many studies that show the mutation profile of radiation-induced tumors but it still needs to be examined for more tumor types, sites, and related factors.

10.2.4.1 Glycophorin A in MN Blood Group Heterozygotes

GYPA is a glycoprotein present on the surface of erythrocytes. The gene has two allelic forms, gpaM and gpaN, which give rise to proteins and differ only in two amino acid residues. The GYPA assay detects and quantifies erythrocytes with allele-loss phenotypes at the autosomal locus that are responsible for the polymorphic MN blood group. It is efficiently analyzed by a pair of techniques such as allele-specific monoclonal antibodies and flow cytometry technique and is sensitive to >1 Gy dose. From these techniques, it is possible to detect two different variant phenotypes: (1) simple allele loss, and (2) allele loss followed by reduplication of the remaining allele. Both give information about the mechanisms underlying "loss of heterozygosity (having two different alleles of the same gene)" at tumor-suppressor genes.

10.2.4.2 Hypoxanthine-Guanine Phosphoribosyl Transferase Gene

The HPRTase is an enzyme that plays a role in the purine salvage from degraded DNA and is controlled by the HPRT gene. The HPRT genes also catalyze the transformation of purine analogs such as 6-thioguanine (6-TG). 6-TG is cytotoxic to normal cells. Hence, cells with mutations in the HPRT gene cannot transform this analog. HPRT-deficient T-lymphocytes

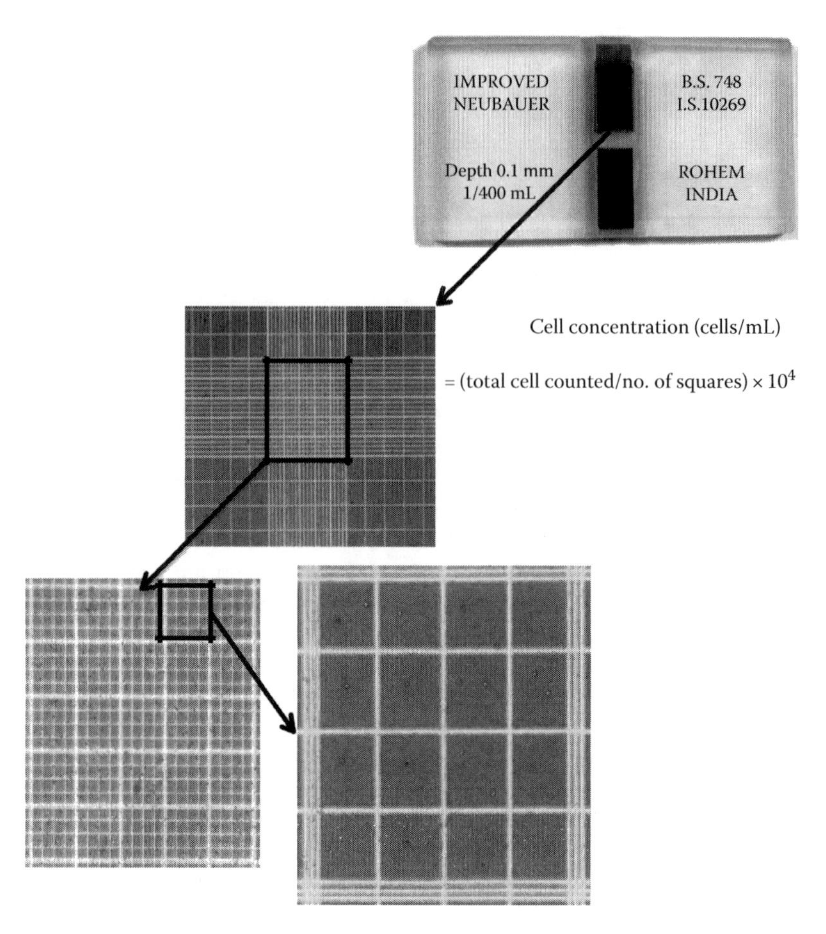

FIGURE 10.1
Cell count by hemocytometer.

determined by the cloning assay are mutant cells resulting from in vivo mutations at the HPRT locus (Albertini, 1999). The HPRT assay is conducted in four steps:

1. Sample blood and isolation of mononuclear cells.
2. Count the cells using a hemocytometer (Figure 10.1). In this method, the sample is mixed with tripan blue dye as given in the protocol. A known quantity of tripan blue mixture is poured into the hemocytometer, which consists of nine squares as seen in Figure 10.1 (cell viability assay). Then the slides are viewed under the microscope. While viewing, the dead cells do not omit foreign substances so they appear blue, unlike the normal cells. The cells in each square are counted using the following formula: multiply the dilution factor by the total number of cells, divide by the number of corner squares counted, and multiply by 10^4 to obtain cell concentration (cells/mL).
3. Plate sufficient number of cells in non-6-TG-containing medium for determining the nonselection cloning efficiency (CE).
4. Plate sufficient number of cells in 6-TG-containing medium for determining the selection cloning efficiency (CE).

5. Incubate all the plates to allow cells to form a colony.
6. Score the number of cells in the positive and negative wells (no growing colonies) under an inverted microscope.
7. Analyze the visible colonies. Visible colonies are considered to be mutants because mutated cells can survive in the presence of this toxic purine analog 6-TG. The mutant frequency (MF) is the ratio between the CE calculated in selective medium and in nonselective medium.

10.2.5 Biomarkers Related to Transcriptional and Translational Changes

As mentioned in Chapter 1, replication is carried out by DNA polymerase in DNA to synthesize DNA from DNA, transcription is carried out by RNA polymerase to synthesize RNA from DNA, translation is carried out by ribosomes to synthesize proteins from RNA, and reverse transcription is carried out to synthesize complimentary cDNA from RNA. So any changes in RNA and protein level due to irradiation may lead to transcriptional and translational changes. Hence, it is used as one of the biomarkers. However, the response to this biomarker on health is not yet clear, which is why further research in this area is encouraged.

10.2.5.1 Biomarkers Related to Changes in RNA Levels

Among many possible techniques, the following are commonly used to analyze transcriptional responses:

1. Microarray-based technology—to analyze the mRNA expression of all known genes simultaneously and to predict early or late development of adverse normal tissue reactions to radiotherapy.
2. Real-time PCR assays—for measuring circulating mRNAs
3. Next-generation sequencing (NGS) technology—low-cost method to sequence entire transcriptomes (RNAseq)
4. Nanostring™—to count biological molecules directly.

10.2.5.2 Biomarkers Related to Changes in Protein Levels

Different organs respond to radiation by altering the level of protein expression and their posttranslational modification status. Therefore, it is believed that protein expression profiling can be used as a biomarker in biological samples such as urine, blood/serum, or even tissue. Protein-rich samples of biological fluids such as serum, urine, or saliva are collected and quantified using high-throughput proteomic technologies, preseparation methods, and mass spectrometry. Since protein expression varies depending on the time, radiation dose, and other related factors, profiling it is a challenging task.

Previous studies show that changes in the level of only two proteins have been used as bio-indicators for radiation exposure. They are (1) amylase, which indicates radiation-induced damage of the parotid gland, and (2) Flt3-ligand, a hematopoietic cytokine indicating damage of the bone marrow. These two proteins are measured from serum/plasma using a clinical blood chemistry analyzer or commercial sandwich ELISA, respectively.

10.2.5.3 Transcriptional and Translational Changes in Cytokines

In addition to the Flt3-ligand protein mentioned earlier, studies proved the involvement of cytokines (low–molecular weight glycoproteins) in response to ionizing radiation. It is known that cytokines (low–molecular weight glycoproteins), produced by immune cells, are responsible for immunity, inflammation, cell–cell communication, cell growth, apoptosis, inhibitors of apoptosis, protein synthesis, ROS generation, activation of antioxidants, and repair processes. It has been shown that there is an involvement of cytokines in response to ionizing radiation. Hence, transcriptional modulation of several cytokines due to irradiation may affect all the previously mentioned responsibilities. For example, at low dose (<1.2 mGy) and low dose rates, cytokines play an important role in the establishment of bystander effects (i.e., response of cells that were not directly irradiated but were in contact with irradiated cells or irradiated cell medium). Microarray experiments and ELISA can be used to analyze the transcriptional change in cytokine. The details of microarray and ELISA techniques are discussed later in this chapter.

10.2.6 Other Biomarkers

In addition to all the biomarkers described, the following are also used to quantify the biological effects of radiation.

10.2.6.1 Epigenetic Biomarkers

Exposure to radiation causes epigenomic (intracellular and extracellular changes and stimuli) modifications, which induce DNA damage and affect gene regulation. This further develops malignancies if the change is on a large scale. These epigenomic modifications include (1) histone modifications, (2) DNA methylation, (3) mRNA binding, and (4) protein phosphorylation. The changes in the profile of these radiation-induced modifications are likely to be dose- and time-dependent, and it remains to be fully established whether the various modifications are interrelated or mutually exclusive.

10.2.6.2 Biomarkers Associated with Cell Cycle Delay, Apoptosis, and Cell Survival

Many bioassays are available to access radiosensitivity based on cell cycle delay, which in turn leads to reduction in cell survival due to irradiation. Regarding apoptosis, significant variation in apoptotic responses have been observed in breast cancer patients and hence it may also be used as a biomarker.

10.2.6.3 Biophysical Markers

When radiation interacts with matter, free radicals are formed. As the dose increases, the number of radicals also increases. The free radicals in solid biological materials such as bones, tooth enamel, finger nails, and hair remain for a sufficiently long time. The concentration of free radicals in these materials is measured by electron paramagnetic resonance (EPR) dosimetry based on the correlation between the intensity or amplitude of the radiation-induced signals and the dose absorbed.

10.3 Various Phases of Biological Dosimetry

A variety of biological samples can be used for measurements with appropriate ethical approvals and informed consent. These include blood, tissues, saliva, buccal cells, skin fibroblasts, urine, feces, hair, hair follicle cells, and nail clippings. The biological dosimetry using these samples consists of three major phases:

1. Sample collection phase
2. Sample processing phase
3. Data analysis phase

10.3.1 Sample Collection Phase

As specified by Holland et al. (2003), the following nine factors should be followed while collecting biological samples for biodosimetry.

10.3.1.1 Factor 1: Communication

It is essential to establish clear communication between scientists, staff, public, and patients for reliable and consistent sample collection and also to minimize additional inconvenience to patients and discomfort to the public.

10.3.1.2 Factor 2: Noninvasive Methods of Sample Collection

It is preferred to adopt less invasive methods such as blood collection, exfoliated cells collection from the mouth (buccal), urine (urothelial), or volatile organic compounds (VOCs) based on the requirement of the study. But invasive sample collection such as biopsy (removal of tissues) is also necessary for specific analyses.

10.3.1.3 Factor 3: Timing

Multiple time points of sample collection are often necessary in order to obtain the true time course of the relationship of the exposure and development of the outcome, and to establish causal associations. For example, levels of mercury in women's hair and blood samples before and during pregnancy may be more informative about the exposure to the embryo than samples collected from the mother after the baby is born. Samples collected a long time before the onset of the disease may be more informative and better associated with the cause of the disease.

Approximately 10 mL of sample (if blood) may be collected within a few hours of whole-body radiation exposure. However, in the case of partial-body or nonuniform exposure, the lymphocytes in the circulating and extravascular pools will not have reached equilibrium until about 24 hours. This could result in an unrepresentative proportion of irradiated cells in the specimen and, therefore, delaying sampling until at least the next day is advisable.

10.3.1.4 Factor 4: Stability of Samples

Factors that affect the stability of biological samples include (1) anticoagulants, (2) stabilizing agents, (3) temperature, (4) timing before initial processing, (5) sterility, (6) endogenous degrading properties (enzymes, cell death), and so on.

10.3.1.5 Anticoagulants

The selection of anticoagulant is very important because certain anticoagulants are better for some analyses and not for others. For example, citrate-stabilized blood may afford better quality of RNA and DNA than other anticoagulants would, and produces a higher yield of lymphocytes for culture, whereas heparin-stabilized blood affects T-cell proliferation and heparin binds to many proteins. Ethylenediaminetetraacetic acid (EDTA) is good for DNA-based assays, but it will influence Mg^{2+} concentration, which will result in poor cell growth and pose problems for cytogenetic analysis (by increasing sister chromatid exchanges, decreasing mitotic index, etc.). The collection of whole blood in any type of anticoagulant-containing tubes may cause the induction of cytokine production in vitro, and likely result in artificially elevated concentrations.

10.3.1.6 Stabilizing Agents

Many components of blood that are potential biomarkers need to be preserved using stabilizing agents. For example, EDTA and ascorbic acid are stabilizing agents for folate in blood, and should be added as soon as possible after blood collection to assure the quality of the analysis. Metaphosphoric acid or reduced glutathione are used to preserve ascorbic acid.

10.3.1.7 Timing Before Initial Processing

The allowable time between collection and processing of biological samples depends on the component(s) of interest and their stability. If high cell viability is desired, processing of blood, buccal swabs, or urine samples should be conducted within 24–48 hours. If the physical distance between the collection and processing facilities involves mailing or transportation delays, analysis of unstable biological end points should be excluded.

10.3.1.8 Temperature

Temperature may affect sample stability at two stages: during the time between sample collection and sample processing and during short- and long-term storage. Ideally, the sample should be separated into different components (plasma, cells) immediately after collection and each component kept at the appropriate temperature.

Generally, isolated DNA is stored at 4°C for several weeks, at −20°C for several months, and at −80°C for several years. Isolated RNA must be stored at −80°C. Live cells are stable at room temperature for up to 48 hours but must be either cultured or cryopreserved in liquid nitrogen at −150°C in order to remain alive. Serum and plasma contain a large number of soluble molecules and most require very low temperature to remain intact (−80°C).

Low temperature (4°C) is often a good compromise between the two extremes of freezing and room temperature: cells can remain viable (compared to room temperature) and it also protects, at least to some extent, against enzymatic degradation of sensitive protein biomarkers.

10.3.1.9 Sterility

Bacteria or fungal contamination can affect the quality of the biomarkers by introducing new products and metabolites. Hence, the requirement for aseptic conditions during the collection process is essential if the intention is to isolate RNA or to culture cells from the sample.

10.3.1.10 Degradation

Enzymatic degradation may affect many biochemical biomarkers. Proteins are sensitive to degradation by proteases. Protein integrity is protected by the addition of commercially available protease inhibitors to the sample immediately after collection and also by processing it on ice. But these protease inhibitors are toxic to live cells, and therefore must not be added to whole blood if cell viability is desired. RNA is also particularly sensitive to degradation by abundant and ubiquitous RNAses and it is secured by adding commercially available RNAse inhibitors but it is not protected at low temperatures.

In contrast, DNA is the most stable component in biological samples, including blood, exfoliated cells, and other tissues. There are reports showing that DNA from exfoliated cell specimens was stable for up to a week at room temperature. The stability of DNA allows it to be retrieved and analyzed from dried bodily fluids, clotted blood, Guthrie cards, dried blood smears on slides, or from clothing, as is often the case in forensic investigations.

10.3.1.11 Factor 5: Containers/Equipment

The choice of the size and characteristics of tubes, bottles, or other containers for sample collection and transportation depends on the sample volume, means of transfer to the laboratory, their cost, storage efficiency, and the type of intended analyses.

Generally, small blood samples can be collected by finger prick on commercially available cards. Buccal cells are commonly collected with a small commercially available mouthwash, simple mouth rinse, cyto-brush, or tongue depressor. Certified RNAse-free containers must be used for handling RNA samples. Sterile single-use containers must also be exclusively used when cells are isolated for culture and/or cryopreservation.

10.3.1.12 Factor 6: Safety

Several issues of safety arise when handling human biological materials since they are infectious and hence precautions must be taken at all stages of work.

10.3.1.13 Factor 7: Shipment

Biological materials pose a very high risk of spreading infectious disease during transportation. Hence, international laws and federal regulations should be followed in terms of packaging, labeling, and documentation of shipped goods according to their classification.

10.3.1.14 Factor 8: Paper Trail

An appropriate paper trail includes collection details, including date, sample number, types and volumes, shipping information, and chain of custody forms. Personal information about the public and patient should be converted into a specific understandable code, in compliance with the privacy protection regulations.

10.3.1.15 Factor 9: Strict Adherence to Protocols

The larger the study, the greater is the challenge for consistency in handling all the biological specimens. Hence, all the operating individuals should be trained to follow all the steps produced in the protocol.

10.3.2 Sample Processing Phase

Whether the original biological sample is whole blood, urine, buccal cells, or other tissue (e.g., biopsies), the processing can produce a variety of specimens for future purposes. The sooner the sample is processed, the better is the quality of the extracted components of interest. More effective sample processing includes provisions for the following:

1. Isolating large quantities of DNA
2. Storing high-quality RNA
3. Using buccal cell DNA, blood clot (or blood smears) for genotyping
4. Separating lymphocyte from granulocyte DNA/RNA
5. Making metaphase spreads (useful for many years)
6. Cryopreserving freshly isolated lymphocytes or whole blood to be recultured
7. Preparing slides of exfoliated cells from mouth and urine

In general, sample processing consists of four major steps: (1) aliquoting (dividing the original sample into many portions); (2) freezing; (3) separating cells from sample; and (4) preparing cells for analysis.

Aliquots prepared for RNA analysis are usually mixed with RNA-stabilizing buffer containing beta-mercaptoethanol (commercially available buffer). Aliquots prepared for analysis of folate are mixed with antioxidant agents (ascorbic acid) and EDTA. Aliquots prepared for immunological biomarkers and comet assay are stabilized in the presence of di-methyl sulfoxide (DMSO).

10.3.2.1 Cryopreservation of Freshly Isolated Cells

Cryopreservation is required to store viable cells that can be recultured in the future to get a larger number of cells because some biomarkers require more cells for analysis. In order to do so, the following processes are carried out:

1. The whole sample (blood) can be cryopreserved in equal volume of a mix of fetal bovine serum (FBS) and DMSO (10% final concentration of DMSO) or DMSO is added directly to whole blood without FBS.
2. Cells are isolated from whole blood before storage and are cryopreserved in FBS/DMSO for future use. Lymphocytes can be isolated from whole blood by density centrifugation through gradients, such as ficoll (a neutral, high-mass, hydrophilic polysaccharide that dissolves readily in aqueous solutions). Differential density properties of the blood components also allow adequate separation by a simple centrifugation step without a density gradient. Lymphocytes and monocytes form a layer (buffy coat) just above the granulocytes and red blood cell pellet.

3. Specific storage conditions are required for successful cryopreservation of cells. Liquid nitrogen in specialized containers (dewars) is used to achieve very low temperatures. Storing the vials in the vapor phase of the liquid nitrogen (−150°C) is also a common practice.
4. As a pilot study, cell viability tests are performed on cryopreserved cells in order to verify the effectiveness of cryopreservation condition, that is, more viable cells and lesser cell loss.

10.3.2.2 Preparation of Cells for Cytogenetic Analysis

Cytogenetic analysis such as MN analysis requires interphase cells but chromosomal aberrations (CA) require metaphase cells. Lymphocytes from blood are the most common cell type used for these kinds of analyses.

For MN analysis, the lymphocyte culture is treated with cell-division inhibitors (cytochalasin B) in order to keep the cell in interphase and then slides are prepared.

Processing cells for CA analysis involves three basic steps: (1) cells in lymphocyte culture can be stimulated to divide by various mitogens (agent that triggers mitosis) including phytohemagglutinin (PHA), concanavalin, interleukin, and pokeweed; (2) to obtain sufficient cells in metaphase, inhibitors of mitosis, such as colcemid, are added 2–4 hours before cell harvest; and (3) the slides are prepared with an optimum density of isolated cells.

All slides for MN and CA analysis can be used immediately or stored for future use. Storage of slides at −20°C in N_2 gas is essential for successful FISH.

10.3.2.3 Preparation of Exfoliated Cells from Mouth and Urine

Exfoliated cells from urine or mouth are easy to obtain. Unlike blood cells, exfoliated cells cannot be as easily grown in culture. For preparation of the exfoliated cells, the cells are washed in tris–HCl/EDTA buffer and then spread on histology slides for use in immediate cytogenetic analysis. Alternatively, they can be frozen at −80°C and serve as a source of DNA in future studies.

10.3.2.4 High-Quality DNA and RNA

There is a higher possibility of nucleic acid damage by degradation enzymes that are released due to the rupture of intracellular organelle membranes and also by apoptosis (programmed cell death). Therefore, for best results, the original sample needs to be handled carefully and under conditions that ensure the cells remain intact. If the cells contain PCR inhibitors, they should also be inactivated by an overnight methanol fixation step in order to maintain the quality of nucleic acids.

High-quality RNA is harder to achieve than high-quality DNA, mostly because of instability. RNA can be extracted effectively from isolated lymphocytes after Ficoll® gradient separation of whole blood. This approach is more cost-effective and produces better RNA yield than direct isolation from whole blood. RNA and DNA purification kits such as Qiagenand and Gentra kits are also commercially available, which include all the necessary buffers and detailed instructions.

Finally, long-term storage (years) may affect the integrity of nucleic acids, if it is not preserved under the right conditions. Both DNA and RNA must be stored at −80°C, although −20°C may be adequate for DNA during shorter periods (months). Multiple aliquots are necessary in order to avoid repeated freezing and thawing and to prevent loss of the entire sample due to cross contamination.

10.3.3 Sample Analysis Phase

In continuation to sample collection and processing, the sample is analyzed using many techniques, some of which are discussed here:

1. Fluorescence in situ hybridization (FISH) technique
2. Comet assay
3. Polymerization chain reaction (PCR)
4. Flow cytometry
5. Western blot
6. Enzyme-linked immunosorbent assay (ELISA)
7. DNA microarray technology

10.3.3.1 Fluorescence In Situ Hybridization Technique

FISH is a cytogenetic technique that perfectly paints chromosomes or portions of chromosomes with fluorescent molecules to detect and localize the presence or absence of specific DNA sequences (gene) on chromosomes. It is a labor-intensive method used to analyze chromosomal aberrations and for gene mapping. FISH uses fluorescent probes (fragment of complementary DNA or RNA to be detected) that bind to only those parts of the chromosome that show a high degree of sequence complementarity. FISH of whole cells is conducted in four steps (Figure 10.2):

1. The sample containing the target cells are fixed. Fixation stabilizes macromolecules and cytoskeletal structures, thus preventing lysis of the cells during hybridization. At the same time, fixation makes the cell permeable (cell and nuclear membranes are opened) for the fluorescently labeled oligo-nucleotide probe molecules.
2. The fixed cells are incubated (to allow hybridization) in a buffer that contains the labeled fluorescent probe at a specified temperature. In general, hybridization is the process of base pairing of two strands of DNA or of RNA, DNA–DNA, DNA–RNA, or with radioactive probes. Ideally, only those probe/rRNA pairs will form, which have no mismatches in the hybrid. Consequently, only target cells that contain the full signature sequence on their rRNA will be stained.
3. The subsequent washing step will remove all unbound probe molecules.
4. Finally, the hybridized cells are detected by epifluorescence microscopy or flow cytometry. Fluorescence microscopy is used to find the location of fluorescent probes in the chromosomes.

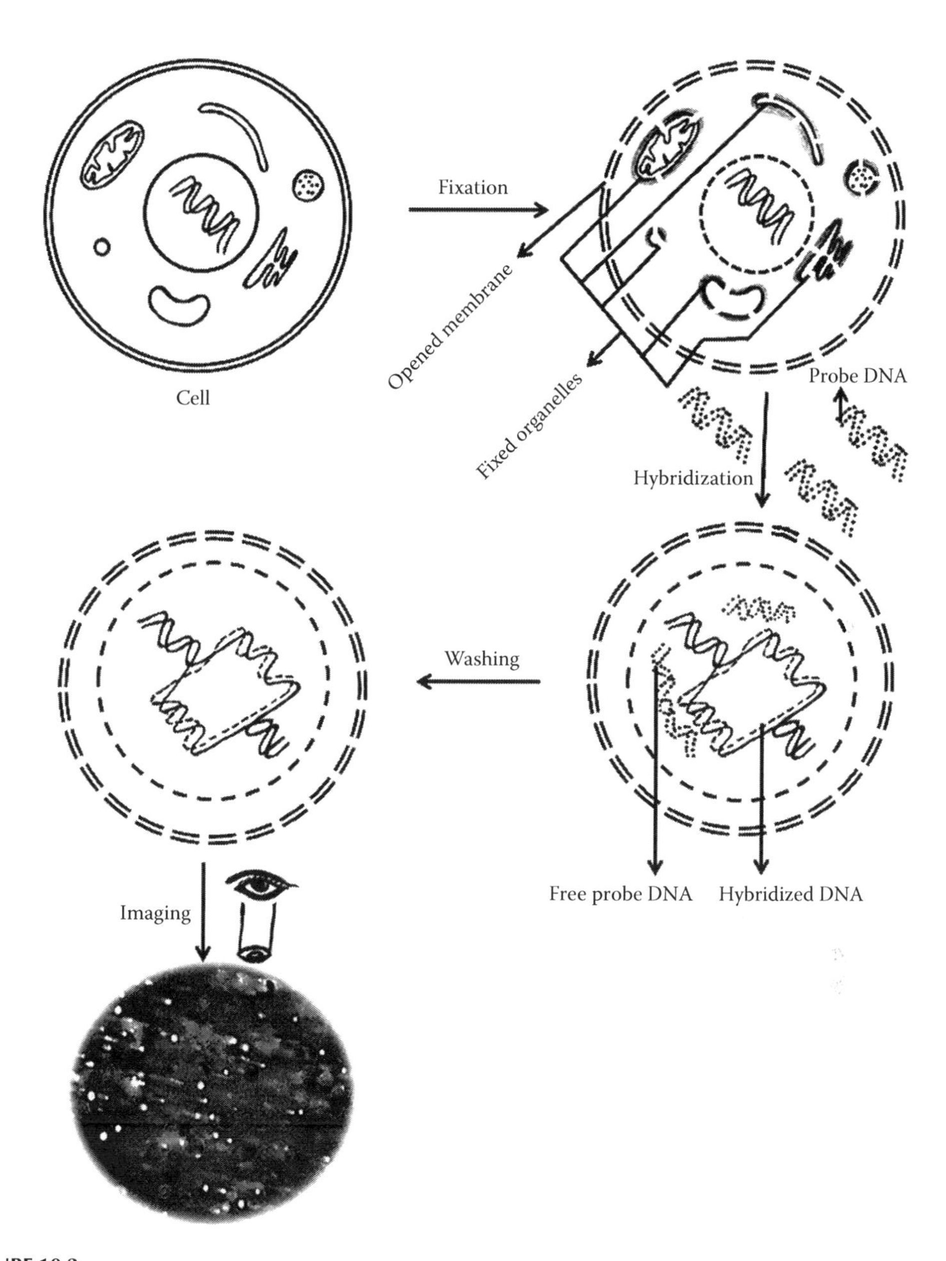

FIGURE 10.2
Schematic representation of FISH technique.

10.3.3.2 Comet Assay

The single-cell gel electrophoresis assay (SCGE, also known as comet assay) is a simple, standard, and sensitive technique that is used to quantify DNA damage (SSB and DSB) and repair. This is possible by detecting the number of nucleotide fragments in the damaged DNA at the level of the individual cells. The following are the general steps involved in a comet assay and these are shown in Figure 10.3.

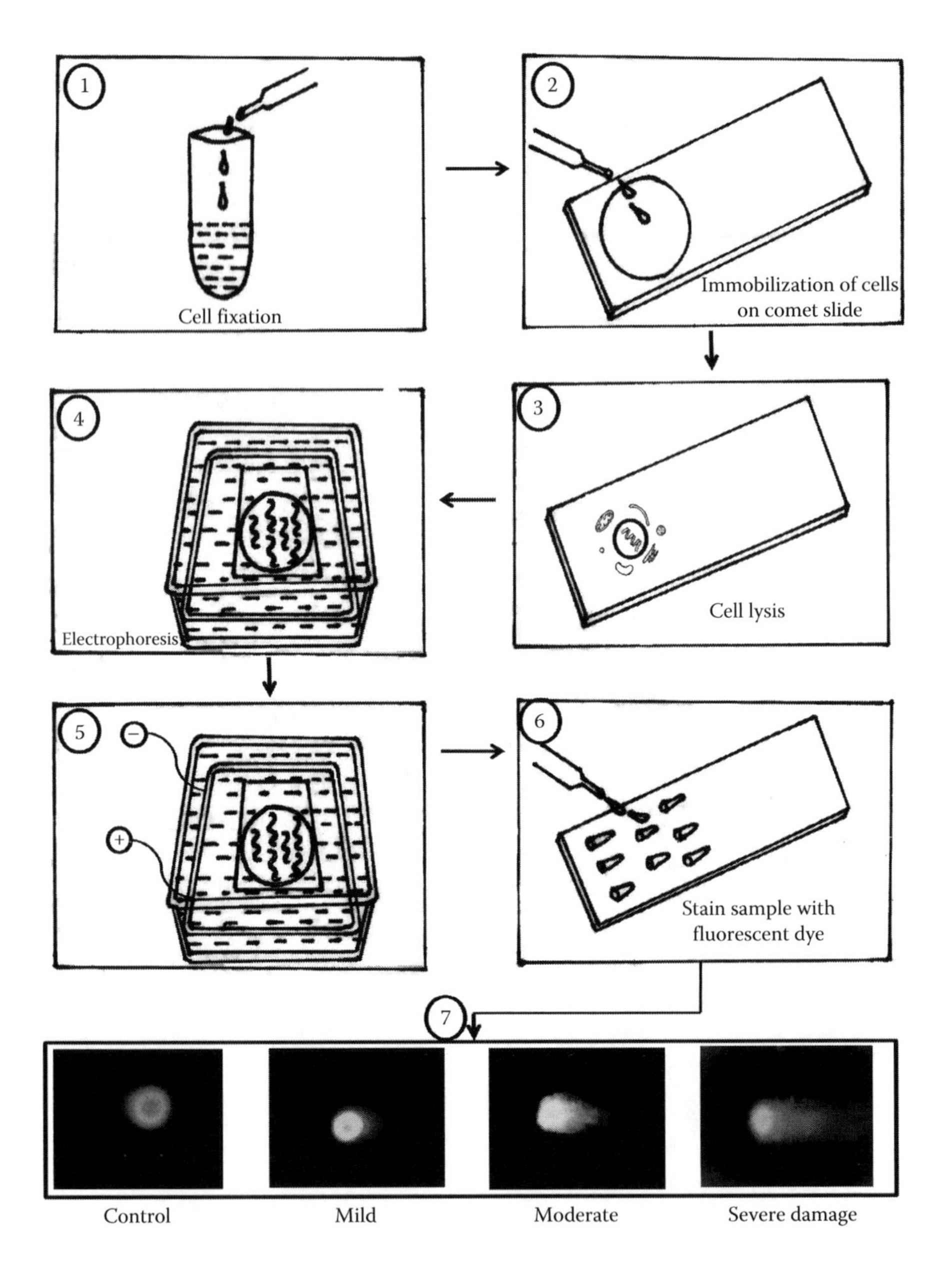

FIGURE 10.3
Comet assay procedure to analyze DNA damage and repair.

1. Mix cells with low melting agarose at 37°C. The agarose forms a matrix of carbohydrate fibers that encapsulate the cells and firmly fix them in place. Therefore, any solution can penetrate the gel and affect the cells without moving it.
2. Immobilize cells on CometSlide.
3. Treat cells with lysis solutions to remove membranes and other soluble cell constituents (like histones) from the DNA.

4. Leave the slides for ~20 minutes in the electrophoresis solution (denaturing alkaline for RNA/neutral for DNA) prior to applying an electric field. In alkaline conditions, the DNA double helix is unwound and denatured, and hence the nucleoid becomes single-stranded.
5. Apply an electric field (typically 1 V/cm) for ~20 minutes. A sample that contains fragments of DNA is forced by an electrical current through the firm gel, which is a screen with small holes of a fixed size. The phosphate group in DNA is negatively charged, so it moves toward a positive electrode with the current. Longer fragments have more nucleotides; therefore, they (1) have a higher molecular weight, (2) are bigger in size, and (3) can't pass through the small holes in the gel and hang there. Shorter fragments are able to pass through and move farther along the gel.
6. Stain the samples with DNA-specific fluorescent dye.
7. Analyze the damage under fluorescence microscopy by combining the amount of DNA in the tail (extension) with the distance of migration using the comet assay software. The numbers of nucleotides in the fragments are also estimated by comparing the fragments to a known sample of DNA fragments that is passed through the gel at the same time.

10.3.3.3 Polymerization Chain Reaction (PCR)

The PCR technology is used to selectively amplify a single or a few copies of a specific DNA piece into thousands or millions of copies. It is applied in genetic testing to know whether or not a particular diseased gene of the parent will be expressed in the child (DNA profiling or parental testing as well): (1) it allows isolation of a particular DNA fragment from genomic DNA by selective amplification, which is possible by the selection of forward and reverse primer; (2) it is used to analyze extremely small amounts of sample; and (3) it permits early diagnosis of malignant diseases. But it is prone to error. Real-time quantitative reverse transcription PCR (qRT-PCR) is an updated PCR technology that enables reliable detection and measurement of products generated during each cycle of the PCR process.

Generally, PCR consists of a series of 20–40 repeated temperature changes, called cycles. Each cycle consists of three discrete temperature steps (increase, decrease, and increase). The temperatures used and the length of time they are applied in each cycle depend on (1) the enzyme used for DNA synthesis, (2) the concentration of primers and dNTPs (nucleotides) used in the reaction, and (3) the melting temperature of the primers. Each cycle is performed in three major steps as shown in Figure 10.4.

1. *Denaturation*: In this step, the DNA samples are heated to ~95°C for 20–30 seconds. This denatures the DNA template by breaking the hydrogen bonds between complementary bases, yielding single-stranded DNA molecules.
2. *Annealing and binding*: The reaction temperature is lowered to ~55°C for 20–40 seconds allowing annealing of both the forward and reverse primers to the single-stranded DNA template. This temperature must be low enough to allow the primer to bind to the strand (hybridization), but high enough to bind the primer perfectly and specifically. Then, the polymerase binds to the primer–template hybrid and begins DNA formation.

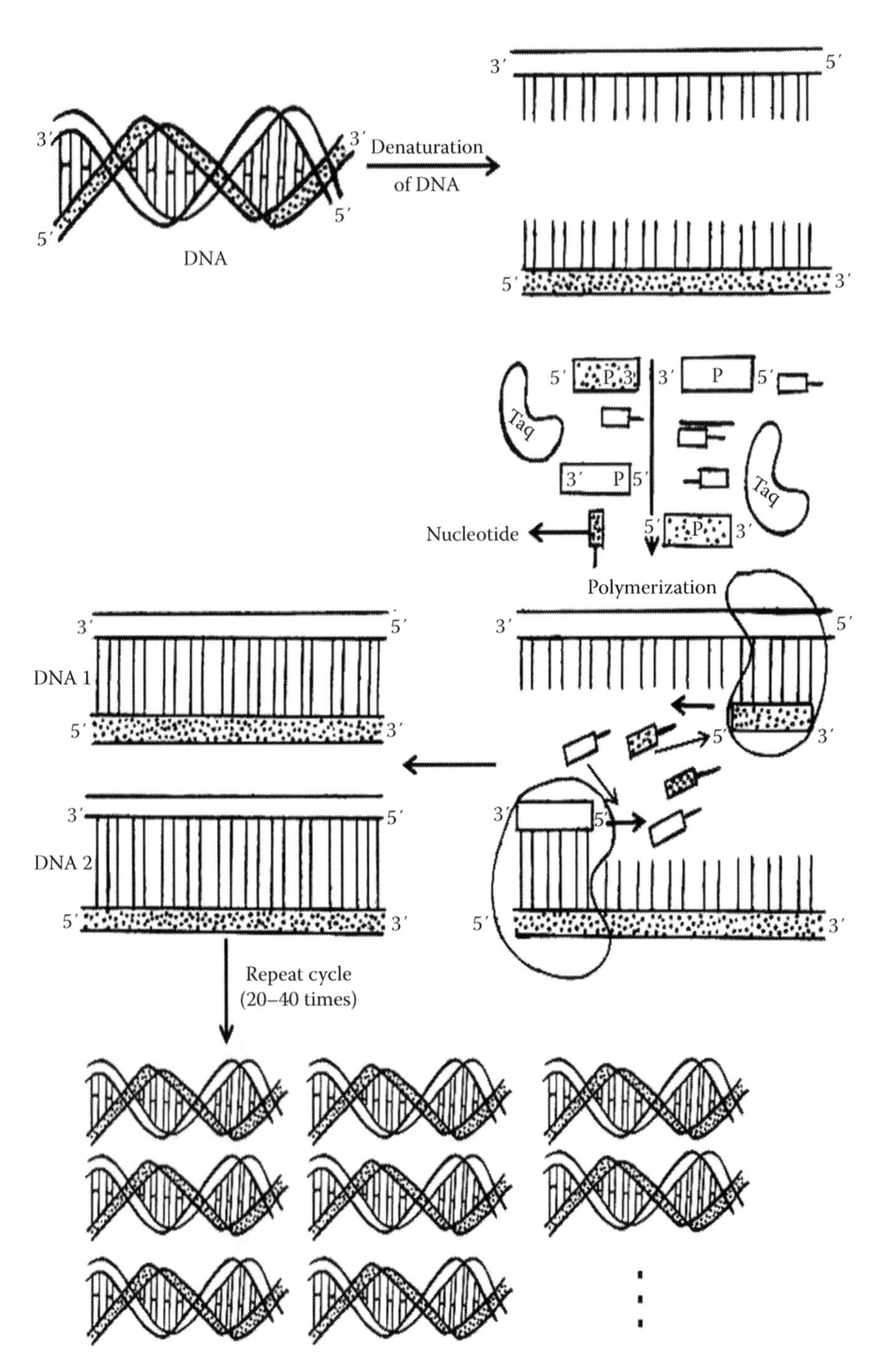

FIGURE 10.4
Schematic representation of PCR.

3. *Extension/elongation*: In this step, the temperature is increased again to ~72°C so that the DNA polymerase synthesizes a new DNA strand complementary to the DNA template strand by adding dNTPs that are complementary to the template. At the end of each cycle, the DNA target is doubled. The DNA polymerase polymerizes a thousand bases per minute, leading to exponential (geometric) amplification of the specific DNA fragment.

10.3.3.4 Flow Cytometry

Flow cytometry is a laser-based, biophysical technology used to detect the types of particles, to count and separate different types of particles in the cell suspension, and for biomarker detection, protein engineering, and, in turn, to detect health disorders. It is similar to a microscopy, except that, instead of producing an image of a cell, flow cytometry offers automated quantification of physical and chemical characteristics of thousands of particles in a second. The basic principle of flow cytometry is the passage of cells in single file in front of a laser so they can be detected, counted, and sorted. It is performed by three main parts of the system (Figure 10.5):

1. *Flow cell*: This carries and aligns the cells so that they pass through the light beam in single file for sensing.
2. *Optical systems*: This consists of lamps, and high-power and low-power lasers.
3. A multiple detector system and other electrical assembly such as analog-to-digital conversion (ADC) and amplifiers. One of these detectors is in line with the light beam and is used to measure forward scatter and another detector is placed perpendicular to the stream and is used to measure side scatter. Since fluorescent labels are used to detect the different cells or components, fluorescent detectors are also in place. The suspended particles or cells, which may range in size from 0.2 to 150 μm, pass through the beam of light and scatter the light beams. The fluorescently labeled cell components are excited by the laser and emit light at a longer wavelength than the light source. This is then detected by the detectors. The detectors therefore pick up a combination of scattered and fluorescent light. These data are then analyzed by a computer that is attached to the flow cytometer using special software.

10.3.3.5 Western Blot

The **western blot** (or **protein immunoblot**) is a widely used technique to detect a specific protein from a complex mixture of proteins in a tissue sample, extract that protein, and finally quantify its expression. This technique uses five major steps to perform the task (Figure 10.6):

1. *Sample preparation*: After diluting with a loading buffer, heat the sample in order to denature the higher-order structure, while retaining the sulfide bridges and the negative charge of amino acids.
2. *Gel electrophoresis*: The proteins (antigen sample) have a negative charge (as they have been denatured by heating) loaded on the gel and are allowed to travel toward the positive electrode when the voltage is applied.

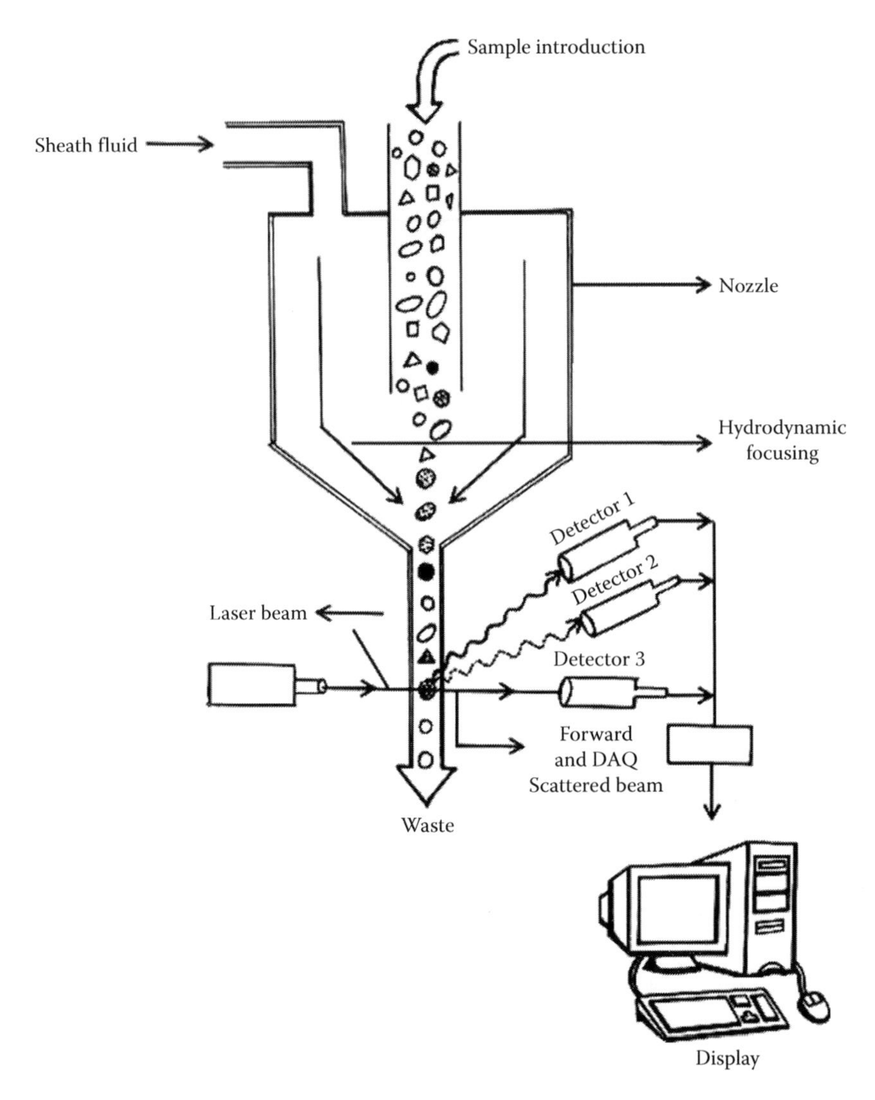

FIGURE 10.5
Schematic representation of a flow cytometer.

3. *Blotting*: The proteins from the gel are then transferred to a membrane (typically nitrocellulose) that produces a band for each protein. The transfer is done using an electric field oriented perpendicular to the surface of the gel, causing proteins to move out of the gel onto the membrane. This type of transfer is called electrophoretic transfer.
4. *Blocking, washing, and antibody incubation*: Blocking agents are added to prevent antibodies from binding to the membrane nonspecifically. After blocking, a primary antibody that is specific to the protein is added, incubated, and washed in a washing buffer, namely phosphate-buffered saline (PBS), to remove unbound antibodies. Then, secondary antibodies (species-specific portion of the primary antibody) are added, incubated, and washed.

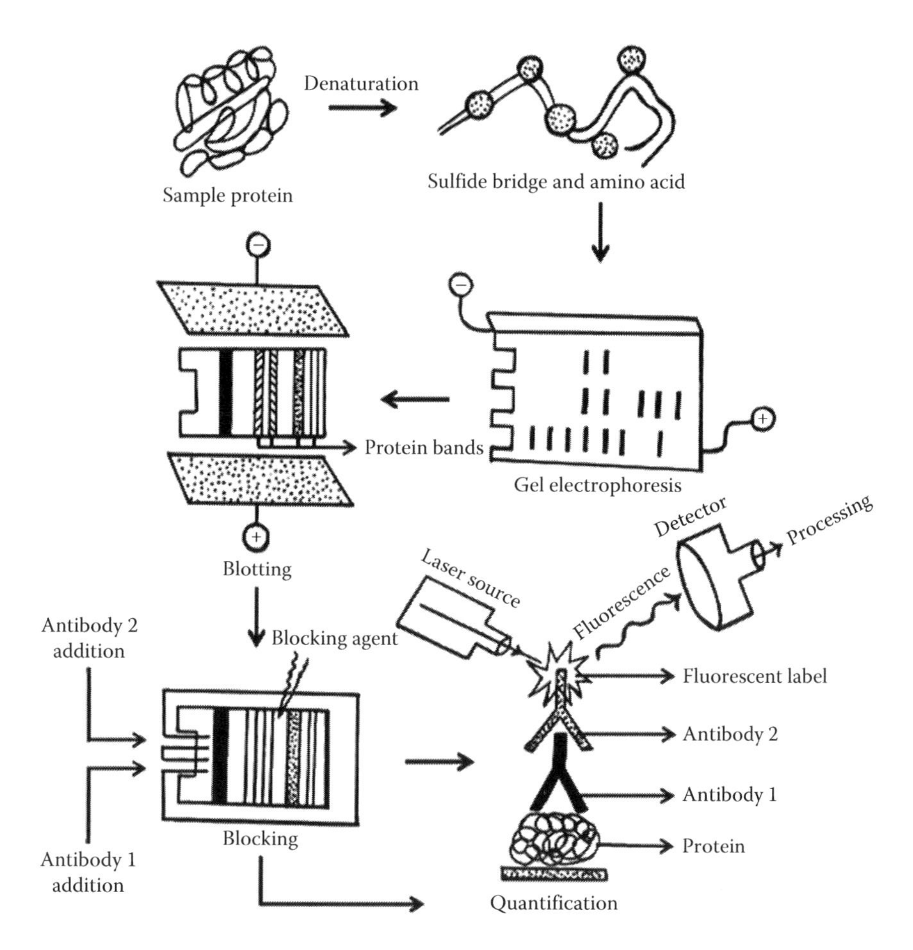

FIGURE 10.6
Major steps involved in western blot procedure.

5. *Quantification*: The bound antibodies (bind on the proteins) are quantified either by developing the film or by fluorescent technique after the addition of a fluorescent label. As the antibodies only bind to the protein (antigen) of interest, only one band should be visible under the fluorescence microscope. The thickness of the band represents the amount of that particular protein present. The analysis requires attention as it is not an absolute measure of quantity due to variation in transfer rate between samples and nonlinear characteristics of the signal generated by the detector with respect to the concentration of the samples (Mahmood and Ping-Chang Yang, 2012).

10.3.3.6 Enzyme-Linked Immunosorbent Assay (ELISA)

ELISA/enzyme immunoassay (EIA) uses the basic immunology concept of an antigen binding to its specific antibody. It is used to detect very small quantities of specific antigens such

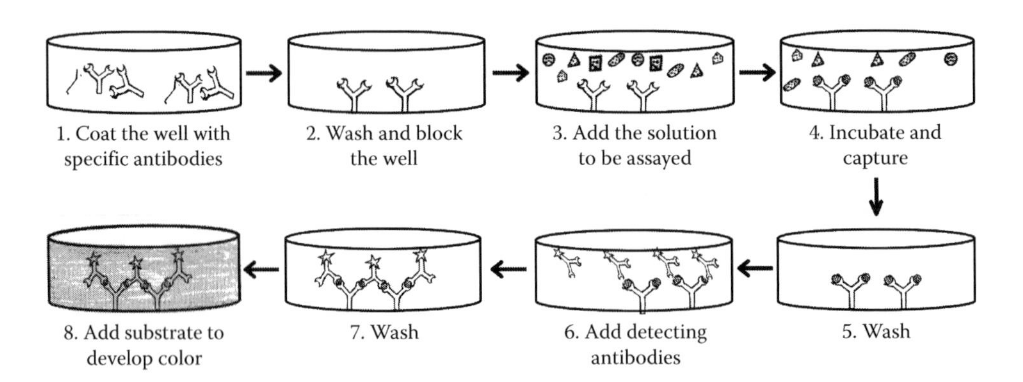

FIGURE 10.7
Simplified diagram of ELISA technique.

as proteins, peptides, hormones, or antibodies that are present in the fluid sample. Among many methods, the sandwiched ELISA is a simple technique as shown in Figure 10.7.

1. A known quantity of antibody is added on the plate surface to capture the desired antigen in the sample and incubated to allow the antibody to bind onto the plate surface.
2. The nonbinding/nonspecific antigens are washed away and the place where there is no antigen is also blocked using bovine serum albumins.
3. Sample blood that contains the target antigen is added on the surface of the plate.
4. This is incubated to allow the target antigen to bind with the antibody.
5. The nonbinding antigens are washed away.
6. Enzyme-linked secondary antibodies (can link to the target antigen-specific enzyme) are added. Since they are specific to that particular antigen, they can link to the antigens. This can also produce chromogenic or fluorescent signal or electrochemical signal when attached to some specific substrate.
7. This is again incubated to allow the secondary antibody to bind with the antigen and the plate is washed to remove all the unbound secondary antibodies.
8. A colorless chemical substrate is added to the medium that binds to the enzyme-linked secondary antibodies. Then, it is converted into a colored substance that can emit a fluorescent or electrochemical signal by the action of that enzyme. This signal is proportional to the amount of target antigen present in the original sample added to the plate.

10.3.3.7 DNA Microarray Technology

DNA microarray technology is an advanced technology used to determine which genes are active and which genes are inactive in different cells. It measures all the genes simultaneously. Microarray technology will help researchers to learn more about many different diseases such as cancer and heart disease. Nowadays, different types of cancers are classified based on the organs in which the tumor develops. However, microarray technology creates an insight into the classification, diagnosis, and treatment of cancers based on the patterns of gene activity in the tumor cells.

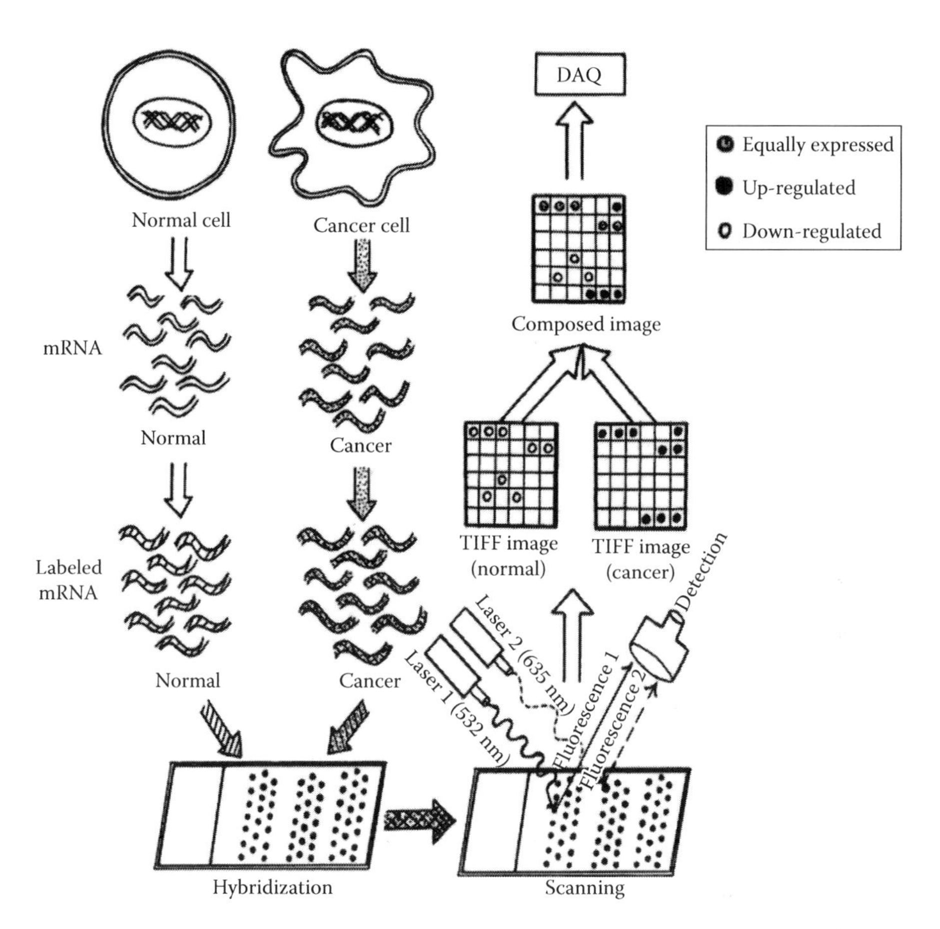

FIGURE 10.8
Pictorial representation of DNA microarray technology.

DNA microarrays are created by robotic machines. It is made up of thousands of spots. This process of analyzing gene expression is performed in seven major steps as shown in Figure 10.8:

1. Collect both normal and cancerous tissues.
2. *Isolate RNA from tissue*: The RNA present in the tissue sample is extracted by (a) dissolving the tissue in a mixture of various organic solvents so that the tissue will dissolve in the solvent and the RNA will be released and (b) centrifuging the sample so that the RNA will float on the top (lesser molecular weight) and other proteins, DNA, and other molecules will settle at the bottom. Now, the RNA is separated from the other contents.
3. *Isolate mRNA from RNA pool*: The separated RNA consists of mRNA, tRNA, and rRNA. But the experiment requires only mRNA molecules because they reflect the gene expression and also have a sequence of adenines known as poly-A tail.

Hence, the solution, which contains all kinds of RNA molecules, is washed over the columns and filled with small beads. The beads bind only to the RNA strands that have a poly-A tail and the other molecules (tRNA and rRNA) are washed out. Then the beads that are attached to the mRNA are removed by further washing with a buffer solution. Likewise, mRNA from both normal and cancer cells may be isolated.

4. *Make labeled cDNA copy by reverse transcription*: mRNA is added with labeling mix to initiate reverse transcription. In general, reverse transcription is referred to as the process of synthesizing single-stranded DNA (complementary DNA or cDNA) using single-stranded RNA as a template and labeling mix.

   The labeling mix consists of (1) poly-T primer, which is attached to the poly-A tails of the mRNA; (2) reverse transcriptase, an enzyme that synthesizes a strand of complementary DNA (cDNA) from the point of primer attachment; and (3) nucleotides (form the new cDNA), which are labeled with green fluorescent dye in the case of normal cells but labeled with red fluorescent dye in the case of cancer cells.

   Then the reverse polymerase assembles the labeled (colored) nucleotides to form the cDNA molecule and the mRNA molecule is ejected. So now there are green-labeled normal cDNA and red-labeled cancer cDNA but there is no mRNA.

5. *Conduct hybridization*: This is the most important step and it is possible by DNA complementary base pairing. In the created microarray plate, each spot (well) contains multiple copies of single genes as DNA probe (single-stranded known DNA sequence). Likewise, if there are approximately 384 spots, it would be possible analyze 384 genes at a time.

   Now, the labeled single-stranded cDNA (lesser than 200 μ volume) prepared from normal and cancer cells is added to the microarray DNA probe. Most of the labeled cDNA molecules hybridize to their microarray DNA probe. A few strands of cDNA molecules may not be hybridized. These nonhybridized molecules are removed by washing the microarray plate.

6. *Scan the microarray*: The hybridized microarray plate is scanned using a microarray scanner, which consists of two laser beams at different wavelengths. At first, it is scanned to find the green-labeled DNA from the healthy cells. In this case, many of the spots are green, but not all. The dark spots are genes that were not transcribed in the healthy cell (not every gene gets expressed in every cell type). Second, it is scanned to find the red-labeled DNA from the cancer cells. In this case, there are many spots that appear red but the pattern is different from the green spot. This means that the cancer cells show different gene expression from normal cells. To get clear information, these two green and red pattern signals are merged together.

7. *Analyze the data*: While merging both the green and red images, mostly yellow signals are captured. The yellow spot (dotted circle in the figure) refers to the combination of both green (empty circle) and red (solid circle) signals. This is not very important because it shows the gene present in both normal and cancer cells. But the red and green signals require special attention. The red spots show the genes that are produced abnormally only in cancer cells by the production of more mRNA through "upregulation" and the green spots show the genes whose normal expression is "downregulation" in the cancer cell.

Objective-Type Questions

1. The best indication for acute radiation sickness is
   a. Nausea
   b. Vomiting
   c. Diarrhea
   d. All
2. The following is a cytogenetic biomarker
   a. Dicentrics
   b. Translocations
   c. Telomere length
   d. All
3. Dicentrics can analyze homogeneous whole-body exposure in the range of
   a. 0.1–5 Gy
   b. 0.25–4 Gy
   c. >5 Gy
   d. <0.1 Gy
4. The ends of linear chromosomes are called
   a. Micromere
   b. Chromomere
   c. Telomere
   d. Centromere
5. The following technology is used to analyze the mRNA expression of all known genes
   a. Microarray-based technology
   b. Real-time PCR assay
   c. NGS technology
   d. Nanostring™
6. Factors that affect the stability of biological sample are
   a. Anticoagulants
   b. Stabilizing agents
   c. Temperature
   d. All
7. In DNA microarray technology, a green dot represents the
   a. Upregulation in cancer gene
   b. Downregulation in cancer gene
   c. Equally expressed
   d. None

8. The following technique is used to selectively amplify a single or a few copies of a specific DNA piece
   a. Microarray-based technology
   b. Real-time PCR assay
   c. NGS technology
   d. Nanostring™
9. Onset of vomiting after 2 hours suggests a dose between
   a. 1–2 Gy
   b. 2–4 Gy
   c. 4–6 Gy
   d. >6 Gy
10. The following is an epigenomic modification
    a. Histone modification
    b. DNA methylation
    c. Protein phosphorylation
    d. All

References

Allen Li X et al., The use and QA of biologically related models for treatment planning: Short report of the TG-166 of the therapy physics committee of the AAPM. *Med Phys* 39(3): 1386–1409, March 2012.

Albertini M, L'unificazione europea e il potere costituente, in: *Nazionalismo e Federalismo*, Bologna, Italy: Il Mulino, 1999.

Alberts B, Johnson A, Lewis J, Raff M, Roberts K, Walter P, *Molecular Biology of the Cell*, 5th edition. New York: Garland Science, 2007.

Armpilia C, Dale RG, Sandilos P, Vlachos L, Radiobiological modelling of dose-gradient effects in low dose rate, high dose rate and pulsed brachytherapy. *Phys Med Biol* 54: 4399–4411, 2006.

Barendsen GW, Dose fractionation, dose-rate and iso-effect relationships for normal tissue responses. *Int J Radiat Oncol Biol Phys* 8: 1981–1997, 1982.

Bentzen SM, Tucker SL, Quantifying the position and steepness of radiation-dose curves. *Int J Radiat Biol* 71: 531–542, 1997.

Bjork-Eriksson T, West C, Karlsson E, Mercke C, Tumour radiosensitiviy (SF2) is a prognostic factor for local control in head and neck cancers. *Int J Radiat Biol Phys* 46: 13–19, 2000.

Bodgi L, Canet A, Pujo-Menjouet L et al., Mathematical models of radiation action on living cells: From the target theory to the modern approaches. A historical and critical review. *J Theor Biol* 394: 93–201, 2016.

Bourhis J, Overgaard J, Audry H et al., Meta-Analysis of Radiotherapy in Carcinomas of Head and Neck (MARCH) collaborative group. Hyperfractionated or accelerated radiotherapy in head and neck cancer: A meta-analysis. *Lancet* 368: 843–854, 2006.

Brada M, Pijls-Johanesma M, De Ruysscher D, Proton therapy in clinical practice: Current clinical evidence. *J Clin Oncol* 25: 965–970, 2007.

Brahme A, Dosimetric precision requirements in radiation therapy. *Acta Radiol Oncol* 23: 379–391, 1984.

Brahme A, Optimization of stationary and moving beam radiation therapy techniques. *Radiother Oncol* 12(2): 129–140, 1988.

Carabe-Fernandez A, Dale RG, Jones B, The incorporation of the concept of minimum RBE into the linear-quadratic model and the potential for improved radiobiological analysis of high-LET treatments. *Int J Radiat Biol* 83: 1–13, 2007.

Carroll QB, *Radiography in the Digital Age: Physics, Exposure, Radiation Biology*. Springfield, IL: Charles C Thomas, 2011.

Coco Martin JM, Mooren E, Ottenheim C et al., Potential of radiation-induced chromosome aberrations to predict radiosensitivity in human tumour cells. *Int J Radiat Biol* 75(9): 1161–1168, 1999.

Cohen L, Theoretical "iso-survival" formulae for fractionated radiation therapy. *Br J Radiol* 41: 522–528, 1968.

Dale RG, The application of the linear quadratic theory to fractionated and protracted radiotherapy. *Br J Radiol* 58: 515–528, 1985.

Dale RG, Radiobiological assessment of permanent implants using tumour repopulation factors in the linear quadratic model. *Br J Radiol* 62: 241–244, 1989.

Dale RG, Coles IP, Deehan C, O'Donoghue JA, Calculation of integrated biological response in brachytherapy. *Int J Radiat Oncol Biol Phys* 38: 633–642, 1997.

Dale RG, Hendry JH, Jones B et al., Practical methods for compensating missed treatment days in radiotherapy, with particular reference to head and neck schedules. *Clin Oncol* 14: 382–393, 2002.

Dale RG, Jones B, The clinical radiobiology of brachytherapy. *Br J Radiol* 61: 153–157, 1998.

Dale RG, Jones B, *Radiobiological Modelling in Radiation Oncology*. London, U.K.: BIR, 2007.

De Santis M, Cesari E, Nobili E, Straface G, Cavaliere AF, Caruso A, Radiation effects on development. *Birth Defects Res C Embryo Today* 81(3): 177–182, September 2007.

Ellis F, Nominal standard dose and the ret. *Br J Radiol* 44: 101–108, 1971.

Ellis HA, Cervical thymic cysts. *Brit J Surg* 54(I): 17–20, 1967.

Evans SM, Freaker D, Hahn SM et al., EF5 binding and clinical outcome in human soft tissue sarcomas. *Int J Radiat Oncol Biol Phys* 64: 922–927, 2006.

Fowler JF, The linear quadratic formula and progress in fractionated radiotherapy. *Br J Radiol* 62: 679–694, 1989.

Garzón O, Plazas MC, Salazar EJ, Evolution of physico-mathematical models in radiobiology and their application in ionizing radiation therapies. *Tecciencia* 9(17): 15–22, 8pp., 2014.

Gregg LS, The hypoxic tumor microenvironment: A driving force for breast cancer progression. *Biochim et Biophys Acta (BBA)—Mol Cell Res* 1893: 382–391, 2016.

Hall EJ, Brenner DJ, The dose-rate effect revisited: Radiobiological considerations of importance in radiotherapy. *Int J Radiat Oncol Biol Phys* 21: 1403–1414, 1991.

Hall EJ, Giaccia AJ, *Radiobiology for the Radiologist*, 6th edition. Philadelphia, PA: Lippincottt Williams & Wilkins, 2006.

Hall EJ, Wuu CS, Radiation induced second cancers: The impact of 3D-CRT and IMRT. *Int J Radiat Oncol Biol Phys* 56: 83–88, 2003.

Han L, Shi S, Gong T, Zhang Z, Sun X, Cancer stem cells: Therapeutic and perspectives in cancer therapy. *Acta Pharm Sin B* 3(2): 65–75, 2013.

Hanahan D, Weinberg RA, Hallmarks of cancer: The next generation. *Cell* 144(5): 646–674, 2011.

Haustermans K, Fowler J, Geboes K et al., Relationship between potential doubling time (Tpot), labeling index and duration of DNA synthesis in 60 esophageal and 35 breast tumors: Is it worthwhile to measure Tpot? *Radiother Oncol* 46: 157–167, 1998.

Hockel M, Knoop C, Schlenger K et al., Intratumoral pO_2 predicts survival in advanced cancer of the uterine cervix. *Radiother Oncol* 26: 45–50, 1993.

Holland NT, Smith MT, Eskenazi B et al., Biological sample collection and processing for molecular epidemiological studies. *Mutat Res* 543: 217–234, 2003.

Holte H, Suo Z, Smeland EB, Kvaloy S, Langholm R, Stokke T, Prognostic value of lymphoma-specific-S-phase fraction compared with that of other cell proliferation markers. *Acta Oncol* 38: 495–503, 1999.

Horriot JC, Bontemps P, van der Bogaert W et al., Accelerated fractionation compared to conventional fractionation improves local-regional control in the radiotherapy of head and neck cancers: Results of the EORTC 22851 randomized trial. *Radiother Oncol* 44: 111–121, 1997.

International Commission and Radiological Protection, Nonstochastic effects of ionizing radiation. ICRP Publication 41. *Ann ICRP* 14(3), 1984. Oxford, U.K.: Pergamon Press.

International Commission and Radiological Protection, Principles of monitoring for the radiation protection of the population. ICRP Publication 43. *Ann ICRP* 15(1), 1985. Oxford, U.K.: Pergamon Press.

International Commission and Radiological Protection, Recommendation. Annals of the ICRP Publication 60. Oxford, U.K.: Pergamon Press, 1991.

Jansen RL, Hupperets PS, Arrends JW et al., MIB-1 labelling index is an independent prognostic marker in primary breast cancer. *Br J Cancer* 78: 460–465, 1998.

Jones B, Dale RG, The potential for mathematical modelling in the assessment of the radiation dose equivalent for cytotoxic chemotherapy given concomitantly with radiotherapy. *Br J Radiol* 78: 939–944, 2005.

Jones B, Dale RG, Deehan C, Hopkins K, Morgan DAL, The role of Biologically Effective Dose (BED). *Clin Oncol* 13: 71–81, 2001.

Jones J, Krag SS, Betenbaugh MJ, Controlling N-linked glycan site occupancy. *Biochim Biophys Acta* 1726(2): 121–137, 2005.

Jones LC, Hoban PW, Treatment plan comparison using equivalent uniform biologically effective dose (EUBED). *Phys Med Biol* 45(1): 159–170, 2000.

Journy N, Laurier D, Bernie M-O, Are the studies on cancer risk from CT scans biased by indication? Elements of answer from a large-scale cohort study in France. *Br J Cancer* 112: 185–193, 2015.

Kaanders JH, Wijffels KI, Marres HA et al., Pimonidazole binding and tumour vascularity predict for treatment outcome in head and neck cancer. *Cancer Res* 62: 7066–7074, 2002.

Kelsey CA, Heintz PH, Chambers GD, Sandoval DJ, Adolphi NL, Paffett KS, *Radiotherapy Treatment Planning: Linear-Quadratic Radiobiology*. Hoboken, NJ: John Wiley & Sons, 2014.

Klokov D et al., Phosphorylated histone H2AX in radiation to cell survival in tumor cells and senografts exposed to single and fractionated dose of x-rays. *Radiother Oncol* 80(2): 223–229, 2006.

Krause C, Rosewich H, Thanos M, Gärtner J, Identification of novel mutations in PEX2, PEX6, PEX10, PEX12, and PEX13 in Zellweger spectrum patients. *Hum Mutat* 27(11): 1157, 2006.

Larsen PR, Conard RA, Knudsen KD et al., Thyroid hypofunction after exposure to fallout from a hydrogen bomb explosion. *JAMA* 247: 1571–1575, 1982.

Law MP, Radiation induced vascular injury and its relation to late effects in normal tissue. *Adv Radiat Biol* 9: 37–73, 1981.

Lea DE, *Actions of Radiation on Living Cells*. Cambridge, U.K.: University Press, 1946.

Lee SP, Leu MY, Smathers JB et al., Biologically effective dose distribution based on the linear quadratic model and its clinical relevance. *Int J Radiat Oncol Biol Phys* 33: 375–389, 1995.

Lehr CM (ed.), *Cell Culture Models of Biological Barriers*. London, U.K.: Taylor & Francis, 2002.

Lehnert S, *Radio Sensitizers and Radio Chemotherapy in the Treatment of Cancer*. Boca Raton, FL: CRC Publisher, p. 25, 2015.

Lemos-Pinto MMP, Cadena M, Santos N, Fernandes TS, Borges E, Amaral A, A dose-response curve for biodosimetry from a 6 MV electron linear accelerator. *Braz J Med Biol Res* 48(10): 908–914, 2015.

Lyman JT, Complication probability as assessed from dose-volume histograms. *Radiat Res Suppl* 8: S13–S19, 1985.

Lynch M, Mutation and human exceptionalism: Our future genetic load. *Genetics* 202(3): 869–875, 2016.

Mahmood T, Ping-Chang Yang N, Western blot: Technique, theory, and trouble shooting. *Am J Med Sci* 4(9): 429–434, 2012.

McParland BJ, *Nuclear Medicine Radiation Dosimetry*. Berlin, Germany: Springer, 2010.

Moiseenko V, Battista J, Van Dyke, J, Normal tissue complication probabilities: Dependence on choice of biological model and dose-volume histogram reduction scheme. *Int J Radiat Oncol Biol Phys* 46: 986–993, 2000.

Moiseenko V, Duzenli C, Durand R, In vitro study of cell survival following dynamic MLC intensity-modulated radiation therapy dose delivery. *Med Phys* 34: 1514–1520, 2007.

Muren LP, Jebsen N, Gustafsson A, Dahl O, Can dose-response models predict reliable normal tissue complication probabilities in radical radiotherapy of urinary bladder cancer? The impact of alternative radiation tolerance models and parameters. *Int J Radiat Oncol Biol Phys* 50(3): 627–637, 2001.

National Research Council (U.S.), Committee on the Biological Effects of Ionizing Radiations, *Health Effects of Exposure to Low Levels of Ionizing Radiation*. BEIR V. Washington, DC: National Academy Press, 1990.

National Research Council (Estados Unidos), Committee to Assess Health Risks from Exposure to Low Level of Ionizing, *Health Effects of Exposure to Low Levels of Ionizing Radiation*. BEIR VII. Washington, DC: National Academy Press, 2006.

Niemerko A, Reporting and analysing dose distributions: A concept of equivalent uniform dose. *Med Phys* 24: 103–110, 1997.

Niemerko A, A generalised concept of equivalent uniform dose (EUD). *Med Phys* 26: 1100, 1999.

Nordsmark M, Bentzen SM, Rudat V et al., Prognostic value of tumour oxygenation in 397 head and neck tumours after primary radiation therapy. An international multi-center study. *Radiother Oncol* 77: 18–24, 2005.

Orton CG, Ellis F, A simplification in the use of the NSD concept in practical radiotherapy. *Br J Radiol* 46: 529–537, July 1973.

Pernot E, Hall J, Baatout S et al., Ionizing radiation biomarkers for potential use in epidemiological studies. *Mutat Res* 751: 258–286, 2012.

Pertea M, Salzberg SL, Between a chicken and a grape: Estimating the number of human genes. *Genome Biol* 11(5): 206, 2010.

Plataniotis GA, Dale RG, Radiobiological modelling of cytoprotection effects of radiotherapy. *Int J Radiat Oncol Biol Phys* 68: 236–242, 2007.

Qi XS, White J, Li XA, Is a/b for breast cancer really low? *Radiother Oncol* 100: 282–288, 2011.

Raleigh JA, Chou SC, Calkins Adams DP et al., A clinical study of hypoxia and metallothionein protein expression in squamous cell carcinomas. *Clin Cancer Res* 6: 855–862, 2000.

Ramsey SD, Zeliadt SB, Richardson LC, Discontinuation of radiation treatment among medicaid-enrolled women with local and regional stage breast cancer. *Breast J* 16: 20–27, 2010.

Reya T, Morrison SJ, Clarke MF, Weissman IL, Review article—Stem cells, cancer, and cancer stem cells. *Nature* 414: 105–111, 2001.

Rodrigues G, Lock M, D'Souza D, Yu E, Van Dyk J, Prediction of radiation pneumonitis by dose—Volume histogram parameters in lung cancer—A systematic review. *Radiother Oncol* 71(2): 127–138, 2004.

Rubin P, Casarett GW, Clinical radiation pathology as applied to curative radiotherapy. *Cancer* 22(4): 767–778, 1968.

Russell WL, The effect of radiation dose rate and fractionation on mutation in mice, in: Sobels F (ed.), *Repair from Genetic Radiation Damage*. Oxford, U.K.: Pergamon Press, pp. 205–217, 231–235, 1963.

Sanchez-Nieto B, Nahum AE, The delta-TCP concept: A clinically useful measure of tumor control probability. *Int J Radiat Oncol Biol Phys* 44(2): 369–380, 1999.

Sanpaolo P, Barbieri P, Genovesi D, Fusco V, Cefaro GA, Biologically effective dose and breast cancer conservative treatment: Is duration of radiation therapy really important? *Breast Cancer Res Treat* 134: 81–87, 2012.

Saunders M, Dische S, Barett A et al., Continous, hypefractionated, accelerated radiotherapy (CHART) versus conventional radiotherapy in small cell lung cancer: Nature data from the randomized multicenter trial. CHART Steering Committee. *Radiother Oncol* 52: 137–148, 1999.

Stewart FA et al., ICRP statement on tissue reactions/early and late effects of radiation in normal tissues and organs—Threshold doses for tissue reactions in a radiation protection context. ICRP Publication 118. *Ann ICRP* 41(1/2): 1–322, 2012.

Strandquist M, A study of the cumulative effects of fractionated X-ray treatment based on the experience gained at the radiumhemmet with the treatment of 280 cases of carcinoma of the skin and lip. *Acta Radiol* 55(Suppl): 300–304, 1944.

Subarsky P, Hill RP, The hypoxic tumour microenvironment and metastatic progression. *Clin Exp Metastasis* 20: 237–250, 2003.

Thames HD, An 'incomplete-repair' model for survival after fractionated and continuous irradiations. *Int J Radiat Biol* 47: 319–339, 1985.

Thames HD, Zhang M, Tucker SL, Liu HH, Dong L, Mohan R, Cluster models of dose-volume effects. *Int J Radiat Oncol Biol Phys* 59: 1491–1504, 2004.

Tomé WA, Fowler JF, Selective boosting of tumor subvolumes. *Int J Radiat Oncol Biol Phys* 48: 593–599, 2000.

Tomé WA, Fowler JF, On cold spots in tumor subvolumes. *Med Phys* 29(7): 1590–1598, 2002.

Torres-Roca JF, Eschrich S, Zhao H et al., Prediction of radiation sensitivity using a gene expression classifier. *Cancer Res* 65: 7169–7276, 2005.

Tucker SL, Zhang M, Dong L, Mohan R, Kuban D, Thames HD, Cluster model analysis of late rectal bleeding after IMRT of prostate cancer: A case-control study. *Int J Radiat Oncol Biol Phys* 64(4): 1255–1264, 2006.

Report of the United Nations Scientific Committee on the Effects of Atomic Radiation, SUPPLEMENT No. 17 (A/3838). Hereditary effects of radiation, UNSCEAR 2001 Report to the General Assembly, with Scientific Annex, 1958.

UNSCEAR, Report of the United Nations Scientific Committee on the effects of atomic radiation, 2001.

Vaupel P, Schlenger K, Knoop C, Hockel M, Oxygenation of human tumours: Evaluation of tissue oxygen distribution in breast cancer by computerized O_2 tension measurements. *Cancer Res* 51: 3316–3322, 1991.

Wang CK, The progress of radiobiological models in modern radiotherapy with emphasis on the uncertainty issue. *Mutat Res* 704(1–3): 175–181, 2010.
West CM, Davidson SE, Roberts SA, Hunter RD, The independence of intrinsic radiosensitivity as a prognostic factor for patient response to radiotherapy of carcinoma of the cervix. *Br J Cancer* 76: 1184–1190, 1997.
West CM, Loncaster JA, Cooper RA et al., Tumour vascularity: A histological measure of angiogenesis and hypoxia. *Cancer Res* 61: 2907–2910, 2001.
West CML, Mj M, Holscher T et al., Molecular markers predicting radiotherapy response: Report and recommendations from an international atomic energy agency technical meeting. *Int J Radiat Oncol Biol Phys* 62(5): 1264–1273, 2005.
Withers HR, Taylor JM, Maciejewski B, The hazard of accelerated tumour clonogen repopulation during radiotherapy. *Acta Oncol* 27: 131–146, 1988.
Wood RD, Mitchell M, Sgouros J, Lindahl T, Human DNA repair genes. *Science* 291(5507): 1284–1289, 2001.
Wyatt RM, Beddoe AH, Dale RG, The effects of delays in radiotherapy treatment on tumour control. *Phys Med Biol* 48: 139–155, 2003.
Wyatt M, Jones BJ, Dale RG, Radiotherapy treatment delays and their influence on tumour control achieved by various fractionation schedules. *Br J Radiol* 81: 549–563, 2008.
Zaider M, Amols HI, A little to a lot or a lot to a little: Is NTCP always minimized in multiport therapy? *Int J Radiat Oncol Biol Phys* 41(4): 945–950, 1998.

Further Reading

Alpen EL, *Radiation Biophysics*. San Diego, CA: Academic Press, 1998.
Albertini RJ, Allen EF, Direct mutagenicity testing in man, in: Walsh PJ, Richmond CR, Copenhaver ED (eds.), *Health Risk Analysis*. Philadelphia, PA, Franklin Institute Press, Proceedings of the Third Life Sciences Symposium, pp. 131–145, 1981.
American Association of Physicists in Medicine, The use and QA of biologically related models for treatment planning. Report of AAPM Task Group 166 of the Therapy Physics Committee, 2012.
Asimov I, Dobzhansky T, *The Genetic Effects of Radiation*. Oak Ridge, TN: U.S. Atomic Energy Commission—Division of Technical Information, 1966.
Awwad H, *Radiation Oncology: Radiobiological and Physiological Perspectives*. Dordrecht, the Netherlands: Springer, 1990.
Bentzen SM, Constine LS, Deasy JO, Eisbruch A, Jackson A, Marks LB, Ten Haken RK, Yorke ED, Quantitative analyses of normal tissue effects in the clinic (QUANTEC): An introduction to the scientific issues. *Int J Radiat Oncol Biol Phys* 76(3 Suppl): S3–S9, March 1, 2010.
Chawapun N, Update on clinical radiobiology. *Biomed Imaging Interv J* 2(1): e22, 2006.
Cockerham LG, Shane BS, *Basic Environmental Toxicology*. Boca Raton, FL: Taylor & Francis Group, LLC, 1994.
Dale R, Jones B, *Radiobiological Modelling in Radiation Oncology*. London, U.K.: The British Institute of Radiology, 2007.
Donald CJ, Nahum AE, *Radiotherapy Treatment Planning: Linear-Quadratic Radiobiology*. Baton Rouge, FL: CRC Press/Taylor & Francis, 2015.
Duncan W, Nias AHW, *Clinical Radiobiology*, 2nd edition. London, U.K.: Churchill Livingstone, 1989.
Elizabeth E, Eric A, Roles for mismatch repair factors in regulating genetic recombination. *Mol Cel Biol* 20(21): 7839–7844, 2000.
Ellis F, Is NSD-TDF useful to radiotherapy? *Int J Radiat Oncol Biol Phys* 11: 1685–1697, 1985.
Forshier S, *Essentials of Radiation Biology and Protection*, 2nd edition. Cengage Learning, 2009.
Hall EJ, Brenner DJ, Cancer risks from diagnostic radiology. *Br J Radiol* 81: 362–378, 2008.

Hall EJ, *Radiobiology for the Radiologist*, 6th edition. Philadelphia, PA: JB Lippincott, Williams & Wilkins, 2005.

Hall EJ, Giaccia AJ, *Radiobiology for the Radiologist*, 7th edition. Lippincott Williams & Wilkins, 2011.

IAEA, *Radiation Biology: A Handbook for Teachers and Students*. Training Course Series No. 42. Vienna, Austria: IAEA, 2010.

International Commission and Radiological Protection, Recommendation. Annals of the ICRP Publication 60. Oxford, U.K.: Pergamon Press, 1990.

International Commission and Radiological Protection, Recommendations of the International Commission on Radiological Protection. ICRP Publication 103. *Ann ICRP* 37: 1–332, 2007.

Joiner MC, Kogel AC, *Basic Clinical Radiobiology*, 4th edition. Boca Raton, FL: Taylor & Francis Group, LLC, 2009.

Kelsey CA, Heintz PH, Chambers GD, Sandoval DJ, Adolphi NL, Paffett KS, *Radiotherapy Treatment Planning: Linear-Quadratic Radiobiology*. Hoboken, NJ: John Wiley & Sons, Inc., 2014.

Kutcher GJ, Burman C, Calculation of complication probability factors for non-uniform normal tissue irradiation: The effective volume method. *Int J Radiat Oncol Biol Phys* 16(6): 1623–1630, 1989.

Kutcher GJ, Burman C, Brewster L, Goitein M, Mohan R, Histogram reduction method for calculating complication probabilities for three-dimensional treatment planning evaluations. *Int J Radiat Oncol Biol Phys* 21(1): 137–146, 1991.

Lawenda BD, *The Effects of Radiation Therapy on the Lymphatic System: Acute and Latent Effects*. Oakland, CA: National Lymphedema Network, 2008.

Levitt SH, Purdy JA, Perez CA, Vijayakumar S (eds.), *Technical Basis of Radiation Therapy Practical Clinical Applications*. Berlin, Germany: Springer-Verlag, 2006.

Lyndon BJ, Houston, Evidence Report: Risk of radiation carcinogenesis- human research program space radiation element- approved for public release. April 7, 2016.

MacDonald SM, DeLaney TF, Loeffler JS, Proton beam radiation therapy. *Cancer Invest* 24: 199–208, 2006.

Mark CH, Polina VS, John MF, Christoph HB, Michael JD, Kenneth BT, Thomas AK, DNA binding by yeast Mlh1 and Pms1: Implications for DNA mismatch repair. *Nucleic Acids Res* 31(8): 2025–2034, 2003.

Marthy H (ed.), *Cellular and Molecular Control of Direct Cell Interactions*, 1985 edition. New York: Springer, May 22, 2013.

Mayles P, Nahum A, Rosenwald JC, *Handbook of Radiotherapy Physics: Theory and Practice*. Boca Raton, FL: CRC Press/Taylor & Francis, 2007.

Mundt AJ, Roeske JC, *Intensity Modulated Radiation Therapy: A Clinical Perspective*. Elsevier, 2005.

Pawlik TM, Keyomarsi K, Role of cell cycle in mediating sensitivity to radiotherapy. *Int J Radiat Oncol Biol Phys* 59(4): 928–942, 2004.

Schlegel WC, Bortfeld T, Grosu A-L (eds.), *New Technologies in Radiation Oncology*. Berlin, Germany: Springer, 2006.

Shirley L, *Biomolecular Action of Ionizing Radiation*. New York: CRC Press/Taylor & Francis, 2007.

Smith C, Wood EJ (eds.), *Energy in Biological Systems (Molecular and Cell Biochemistry)*. Springer, 1991.

Tod WS, *Targeted Radionuclide Therapy*. Lippincott Williams & Wilkins, 2011.

Tubiana M, Dutreics J, Wambersie A, *Introduction to Radiobiology*. London, U.K.: Taylor & Francis Group, LLC, 1990.

Tubiana M, Dutreix J, Wambersie A, *Introduction to Radiobiology*. Boca Raton, FL: Taylor & Francis Group, LLC, 2005.

Valentin J, The 2007 Recommendations of the International Commission on Radiological Protection ICRP Publication 103 Approved by the Commission in March 2007. 37(2–4):1–332, March 2007.

Weinberg RA, *The Biology of Cancer*, 2nd edition. New York: Garland Science, 2013.

Yao D, Dai C, Peng S, Mechanism of the mesenchymal–epithelial transition and its relationship with metastatic tumor formation. *Mol Cancer Res* 9(12): 1608–1620, 2011.

Zaider M, Amols HI, Practical considerations in using calculated healthy-tissue complication probabilities for treatment-plan optimization. *Int J Radiat Oncol Biol Phys* 44(2): 439–447, 1999.

Glossary

Absorbed dose Radiation dose (absorbed dose) is the amount of energy absorbed per unit mass of the irradiated medium/tissue.

Acute radiation syndrome Acute radiation syndrome, or radiation sickness, results from irradiation of majority of the body (mostly whole body) to very high doses of external and highly penetrating radiation (x-ray, gamma ray, neutron, etc.) in a short period of time (seconds to minutes).

Anabolism Anabolism is the process of building and storing large biomolecules (proteins and nucleic acids) from small molecules (amino acids and nucleotides) using the energy generated from catabolism.

Anaphase bridge Anaphase bridge is formed during mitosis when telomeres (ends) of sister chromatids fuse together and fail to completely segregate into their respective daughter cells.

Apoptosis Apoptosis is naturally occurring programmed and targeted cell death.

Autophagy Autophagy is a genetically regulated form of programmed cell death.

Biochemistry Biochemistry is the study of life at molecular level.

Biological dosimetry Biological dosimetry (or biodosimetry) is the measurement of induced biological effects (biomarkers) of radiation in exposed individuals that can be related to the magnitude of radiation dose received.

Carcinogenesis Carcinogenesis is the induction of cancer by genetic mutations which occur due to chemicals, radiation exposure, abnormality in metabolism, and so on.

Catabolism Catabolism is the process of breaking large molecules (mostly carbohydrates and fats) into simple molecules and produces energy by way of cellular respiration and heat.

Cataract Cataract is the opacification of the normally transparent eye lens.

Cell cycle Cell cycle is a sequential cyclical pattern of events by which a cell duplicates its genome, synthesizes other constituents of the cell, and eventually divides into two daughter cells.

Cell synchronization Cell synchronization is a process by which cells at different stages of the cell cycle are brought to the same phase.

Deterministic effects Deterministic effects (nonstochastic) are the effects of radiation on an exposed individual (somatic effect) which have a threshold dose value under which they do not appear.

Doubling dose Doubling dose is the dose required to induce the same number of mutations in a generation as that which arise spontaneously.

Fractionation Fractionation is dividing the total dose to be delivered into several small portions, allowing a time period in between each exposure.

Genetic effects When the genes in the reproductive cells of a person are mutated by ionizing radiation, the damage will pass on to his/her generation/progeny/offspring. Hence, it is referred to as the genetic effects of radiation.

Genetically significant dose (GSD) The genetically significant dose (GSD) is the dose received by the parental gonads during a reproductive lifetime, which starts from conception of the mother to the average age of child birthing (approximately 30 years in humans).

Inorganic molecule An inorganic molecule is a substance that does not contain carbon and hydrogen atoms.

$LD_{50/30}$ $LD_{50/30}$ is the whole-body mean dose which is expected to be lethal in 50% of the normal population within 30 days.

$LD_{50/60}$ $LD_{50/60}$ is the whole-body mean dose which is expected to be lethal in 50% of the normal population within 60 days.

Lethal mutations Lethal mutations are mutations that bring about death before the responsible gene passes its character to the next generation.

Lineal energy (y) Lineal energy is the energy imparted in one event divided by the mean chord length l (target volume) that results from the random interception of the site by a straight line.

Metabolism Metabolism is a network of highly coordinated enzyme-catalyzed chemical reactions within the cells of living organisms.

Metastasis Metastasis is the spreading of carcinogenic cells to remote places. It reduces the curability rate rapidly.

Microdosimetry Microdosimetry is the study of the spatial (360° variation), temporal (spatial distribution with time), and spectral aspects (energy and type of radiation) of the stochastic nature of the energy deposition processes in microscopic structures.

Mutation rate Mutation rate is the measure of the rate at which various types of mutations occur over time.

Mutations If the chromosomes are subjected to more serious structural or chemical changes due to various factors (chemical, radiation, environmental factors, etc.), an individual with entirely new characteristics not seen earlier is produced. These changes are referred to as mutations.

Necrosis Necrosis is the irreversible accidental death of living cells. It occurs when there is not enough blood supply to the tissue either due to irradiation or some other agents such as poisoning.

(NTCP) $TD_{5/5}$ (NTCP) $TD_{5/5}$ denotes the tumor dose that has the probability to produce 5% complications in normal tissues within 5 years after treatment.

(NTCP) $TD_{50/5}$ (NTCP) $TD_{50/5}$ denotes the tumor dose that has the probability to produce 50% complications in normal tissues within 5 years after treatment.

Organic molecule An organic molecule is a substance that contains both carbon and hydrogen.

Radiation-induced bystander effect Radiation-induced bystander effect is a phenomenon in which unirradiated cells exhibit irradiated effects (such as reduction in cell survival, cytogenetic damage, apoptosis enhancement, chromosome aberrations, micronucleation, transformation, mutation, etc.) as a result of molecular signals received from nearby irradiated cells.

Radiation mitigator Radiation mitigator is a chemical agent used to prevent acute normal tissue toxicity by interrupting damage pathways, inducing DNA repair mechanism, and triggering normal cells to repopulate.

Radiolysis In general, water molecules absorb energy, ionizes, and disassociates into free radicals. This process is referred to as radiolysis of water.

Radio protector Radio protector is a chemical agent used to reduce the radiation-induced damage in normal tissues by reducing genetic effects and intracellular or interstitial oxygen pressure.

Radiosensitivity Radiosensitivity is the degree of shrinkage of a tumor/normal tissues following irradiation.

Radiosensitizer Radiosensitizer is a chemical agent used to sensitize the tumor to irradiation in order to increase cell killing.

Relative biologic effectiveness (RBE) Relative biologic effectiveness (RBE) is the comparison of a dose of standard 250 kVp x-ray to a dose of various test radiations that produces the same biological response.

Respiration Respiration is the process of oxidizing food molecules, like glucose, to carbon dioxide and water.

Senescence Senescence is the state of permanent loss of a cell's proliferative capacity due to cell aging.

Somatic effects Somatic effects are the biological effects of radiation seen on the exposed individual during his/her life time.

Specific energy Specific energy is the energy imparted locally in a small volume of mass.

Spontaneous mutations Spontaneous mutations are mutations that take place naturally, mostly due to the complicated mechanism of gene replication, without human interference.

Stochastic effects Stochastic effects (random or nondeterministic or probabilistic effects) do not have a threshold dose value and cannot be avoided.

Suicide bags If lysosomes and peroxisomes burst, they begin to digest the cell's protein, causing cell death. For this reason, they are also called "suicide bags."

Teratogenic effects of radiation Teratogenic effects of radiation are the effects of radiation on the developing embryo or fetus due to in utero irradiation.

Transcription RNA is synthesized in the nucleus using DNA as a template by the action of an enzyme RNA polymerase (protein). This process is known as transcription.

Translation Messenger RNA carries information from DNA to the ribosome (in the protoplasm) for protein synthesis in the cell. This process is known as translation.

Treatment time Treatment time is the overall time taken to deliver the prescribed dose from beginning of the treatment until its completion.

Volume effect The radiation response on normal tissues largely depends on the volume of tissue irradiated in addition to the number of cells killed. This is referred to as the volume effect.

Index

C

I

K

L

M

N

O

P

R

S